T0260807

Deep Learning Approaches to Cloud Security

Scrivener Publishing
100 Cummings Center, Suite 541J
Beverly, MA 01915-6106

Advances in Cyber Security

Series Editors: Rashmi Agrawal and D. Ganesh Gopal

Scope: The purpose of this book series is to present books that are specifically designed to address the critical security challenges in today's computing world including cloud and mobile environments and to discuss mechanisms for defending against those attacks by using classical and modern approaches of cryptography, blockchain and other defense mechanisms. The book series presents some of the state-of-the-art research work in the field of blockchain, cryptography and security in computing and communications. It is a valuable source of knowledge for researchers, engineers, practitioners, graduates, and doctoral students who are working in the field of blockchain, cryptography, network security, and security and privacy issues in the Internet of Things (IoT). It will also be useful for faculty members of graduate schools and universities. The book series provides a comprehensive look at the various facets of cloud security: infrastructure, network, services, compliance and users. It will provide real-world case studies to articulate the real and perceived risks and challenges in deploying and managing services in a cloud infrastructure from a security perspective. The book series will serve as a platform for books dealing with security concerns of decentralized applications (DApps) and smart contracts that operate on an open blockchain. The book series will be a comprehensive and up-to-date reference on information security and assurance. Bringing together the knowledge, skills, techniques, and tools required of IT security professionals, it facilitates the up-to-date understanding required to stay one step ahead of evolving threats, standards, and regulations.

Publishers at Scrivener
Martin Scrivener (martin@scrivenerpublishing.com)
Phillip Carmical (pcarmical@scrivenerpublishing.com)

Deep Learning Approaches to Cloud Security

Edited by

Pramod Singh Rathore
Vishal Dutt
Rashmi Agrawal
Satya Murthy Sasubilli
and
Srinivasa Rao Swarna

Scrivener
Publishing

WILEY

Wiley Global Headquarters
111 River Street, Hoboken, NJ 07030, USA

For details of our global editorial offices, customer services, and more information about Wiley products visit us at www.wiley.com.

Limit of Liability/Disclaimer of Warranty
While the publisher and authors have used their best efforts in preparing this work, they make no representations or warranties with respect to the accuracy or completeness of the contents of this work and specifically disclaim all warranties, including without limitation any implied warranties of merchantability or fitness for a particular purpose. No warranty may be created or extended by sales representatives, written sales materials, or promotional statements for this work. The fact that an organization, website, or product is referred to in this work as a citation and/or potential source of further information does not mean that the publisher and authors endorse the information or services the organization, website, or product may provide or recommendations it may make. This work is sold with the understanding that the publisher is not engaged in rendering professional services. The advice and strategies contained herein may not be suitable for your situation. You should consult with a specialist where appropriate. Neither the publisher nor authors shall be liable for any loss of profit or any other commercial damages, including but not limited to special, incidental, consequential, or other damages. Further, readers should be aware that websites listed in this work may have changed or disappeared between when this work was written and when it is read.

Library of Congress Cataloging-in-Publication Data

ISBN 9781119760528

Cover image: Stockvault.com
Cover design by Russell Richardson

Set in size of 11pt and Minion Pro by Manila Typesetting Company, Makati, Philippines

10 9 8 7 6 5 4 3 2 1

Contents

Foreword

This is Dr. Abhishek Kumar, Assistant Professor in Chitkara University, Himachal Pradesh. I have been involved in the research for more than 8 years with the authors of this book. This book is about a solution to these more intuitive problems. This solution is to allow computers to learn from experience and understand the world in terms of a hierarchy of concepts, with each concept defined through its relation to simpler concepts.

This book is about how Deep Learning is the fastest growing field in computer science. Deep Learning algorithms and techniques are found to be useful in different areas like Automatic Machine Translation, Automatic Handwriting Generation, Visual Recognition, Fraud Detection, Detecting Developmental Delay in Children. However, applying Deep Learning techniques or algorithms successfully in these areas needs a concerted effort, fostering integrative research between experts ranging from diverse disciplines from data science to visualization. This book provides state of the art approaches of Deep Learning in these areas. It includes areas of detection, prediction, as well as future framework development, building service systems and analytical aspects. In all these topics, approaches of Deep Learning such as artificial neural networks, fuzzy logic, genetic algorithm and hybrid mechanisms are used. This book is intended for dealing with modeling and performance prediction of the efficient cloud security systems thereby bringing newer dimension.

This book shall help clarify understanding of certain key mechanism of technology helpful in realizing such system. Enables processing of very large dataset help with precise and comprehensive forecast of risk and delivers recommended action that improve outcome for consumer. It is a novel application domain of deep learning that is of prime importance to human civilization as a whole. This would be helpful for both professionals and students, with state-of-the art knowledge on the frontiers in information assurance. This book is a good step in that direction.

Dr. Abhishek Kumar
Assistant Professor
Abhishek Kumar || Assistant Professor, PhD, Senior Member (IEEE)
Chitkara University Research and Innovation Network (CURIN)
Chitkara University, India

Preface

This book is organized into fifteen chapters. Chapter 1 discusses the prevailing Biometric modalities, classification, and their working. It goes on to discuss the various approaches used for Facial Biometric Identification such as feature selection, extraction, face marking, and the nearest neighbor approach.

In Chapter 2 we understand the cloud computing concept with Multi-Tenant Framework (MWF). In Multi-Tenant Framework, there is a requirement of privacy and security, a concept developed using Deep Learning. The goal is to find privacy requirements in many factors. Multi-tenancy based systems use the Deep Learning concept. The services of Multi-Tenant based systems are aggregated due to the dynamic environment of cloud computing. Three consistencies will maintain privacy policies using deep learning.

In Chapter 3, Automatic Emotion Detection using facial expression recognition is now a main area of focus in various fields such as computer science, medicine, and psychology. Various feature extraction techniques have been developed for classification of expressions from EEG signals from brain and facial expressions from static images, as well as real time videos. Deep Learning plays an important role for this kind of task. This chapter provides a review of research work carried out in the field of facial expression recognition and EEG Signal classification.

In Chapter 4, the main motivation of the proposed work is to improve the efficiency of wind power generation with the use of solar panels and utilize the power generated by solar cells effectively by powering the electrical components used inside the wind mill, such as revolving motor, elevators, etc. In this paper, various literatures have been reviewed and the remarkable features of the proposed system are highlighted. At the end, the data analysis is done using deep learning and all the results are visualized in graphical form. In the analysis, the power generation from our proposed system and traditional methods is visualized.

Chapter 5 discusses mosaicing; a method of assembling multiple over-lapping images of the same scene into a larger, wider image of a scene which overcomes the above issues. In the real-time monitoring, a major problem is that the field of vision is completely too small to capture the target and a larger field of vision results in low resolution. Ability to handle all the above issues, which includes management of quantity and quality feature extraction, is proposed. In this paper, Speeded Up Robust Features (SURF) are used to construct a mosaiced background model for foreground seg-mentation through deep learning.

In Chapter 6, Chronic Kidney Disease shows slow and periodical loss of kidney function over a period of time and will develop into permanent kidney failure when left untreated. The proposed work aims at presenting the use of Deep Learning for the prediction of Chronic Kidney Disease. Training has been performed using 16 attributes of about 400 patients. Three Deep Learning techniques like Random Forest, Support Vector Machine, and the Naive Bayes Classifier are helpful in predicting the stages. A comparative analysis of these three classifiers is performed.

Chapter 7 provides a cognizance into the implementation of a support vector machine algorithm of Deep Learning to diagnose neurological con-ditions. With advancement of emerging technologies, the means of diag-nosing neurological conditions is substantially more complex than it used to be. Procedures and diagnostic tests are tools that help doctors to identify a neurological illness or other medical condition. Precise identification of neurological pathologies can be done by adhering to an autopsy after the person's passing.

In Chapter 8 we define what convolution is needed in neural net-works and the application of pooling on different data sets. This chapter also addresses how to use a CNN and the kind of operations it applies on the data-set. In this chapter, CNN are applied on image assets for feature extraction and dimensionality reduction.

In Chapter 9, we discuss Deep Learning standard confronter issues on which shallow models, like SVM, are highly affected by the menace of dimensionality. As a module of a two-phase learning plan counts numerous layers of non-straight management, a lot of noticeably robust highpoints are logically unglued from the evidence. The existing instructional work-out awarding the ESANN Deep Learning unusual consultation delicacies the cutting-edge models and results in the present understanding of this learning method which is an orientation for some difficult classification activities. So, in this chapter, we will learn about how Cloud Computing and Deep Learning have taken over the world with their new and improved technologies and will learn about their applications, advantages, disadvantages, and correlations regarding different applications.

In Chapter 10, we show a progression of examination concentrates on the best way to quicken the preparation of a disseminated AI Deep Learning model dependent on remote administration. Circulated Deep Learning has become the standard method of present deep learning models preparation. In conventional appropriated deep learning dependent on mass simultaneous equal, the impermanent log jam of any hub in the group will defer the estimation of different hubs in view of the successive event of coordinated hindrances, bringing about, generally speaking, execution debasement. Our paper proposes a heap adjusting methodology named versatile quick reassignment (AdaptQR).

In Chapter 11, we discuss that Deep Learning and big data are the two very emerging technologies. The large amount of data gathered by organizations are used for many purposes like resolving problems in marketing, medical science, technology, national intelligence, etc. In this current world, the old house data processing units are not very efficient to handle, process, and analyze because the collected data is unstructured and very complex. Because of this, deep learning algorithms which are fast and efficient in solving the backlogs of the traditional algorithms are in use now a days. Biometrics uses the pattern recognition technology of Digital Image Processing for identifying unique features in humans. Most commonly applied or considered biometric modalities comprise fingerprint impression, facial landmarks, iris anatomy, speech recognition, hand writing detection, hand geometry recognition, finger vein detection, and signature identification.

In Chapter 12 we discuss that security as an important issue in the cloud and a proper algorithm is needed; with the cloud, security is secure. This method is based on two algorithms: one based on machine learning and the other on a neural system. A machine learning algorithm is based on a KNN algorithm and neural system strategy based on data fragment and hashing technology; both of these processes should optimize cloud security by using cloud data encryption to the cloud server. There are seven applications used for in-depth study described, namely, customer relationship management, image recognition, natural language processing, recommendation programs, automated speech recognition, drug discovery, and toxicology and bioinformatics.

Chapter 13 explains the classification and prediction of network assaults through an algorithm. The real time intrusion detection system finds the network attack using different deep learning algorithms to calculate the accuracy detection rate, false alarm rate, and generate higher accuracy and efficient prediction of network attacks.

In Chapter 14, we discuss that in today's situation, mysterious structures are initiated to empower clients to store and procedurelize their information utilizing distributed computing. These structures are commonly developed utilizing cryptosystems, appropriated frameworks, and, some of the time, a mix of both. To be explicit, homomorphic cryptosystems, Attribute-Based Encryption (ABE), Service-Oriented Architecture (SOA), Secure Multi-Party Computation (SMC), and Secret Share Schemes (SSS) are the significant security systems being gotten to by practically all current usage. The principle issues being looked during the time spent on gigantic information investigation over cloud utilizing these methods are the computational expenses related with all handling errands, violations from other users of the cloud, insufficient security of internet channels, and absence of accessibility of resources.

Pramod Singh Rathore
Assistant Professor,
Aryabhatta College of Engineering and Research Center, Ajmer
Visiting Faculty, Department of Computer Science & Engineering
MDSU Ajmer, India

Dr. Vishal Dutt
Assistant Professor, Aryabhatta College, Ajmer, India|
Visiting Faculty, Department of Computer Science,
MDSU Ajmer, India
vishaldutt53@gmail.com

Professor Rashmi Agrawal
Professor,
Manavrachna International Institute of
Research and Studies, Faridabad, India

Satya Murthy Sasubilli
Solution Architect,
Huntington National Bank

Srinivasa Rao Swarna
Program Manager/Senior Data Architect,
Tata Consultancy Services

Biometric Identification Using Deep Learning for Advance Cloud Security

Navani Siroya[1*] and Manju Mandot[2]

[1]MDS University Ajmer, India
[2]Computer Science, JRN Rajasthan Vidyapeeth University, Udaipur, India

Abstract

A few decades ago, biometric identification was a staple technology of highly advanced security systems in movies, but today, it exists all around us. Biometric technologies have the potential to revolutionize approaches to identity verification worldwide.

This chapter discusses the prevailing Biometric modalities, their classification, and their working. It goes on to discuss the various approaches used for Facial Biometric Identification such as feature selection, extraction, face marking, and the Nearest Neighbor Approach.

Here, we propose a system that compares an input image with that of the database in order to detect the presence of any similarities. Moreover, we use fiducially point analysis to extract facial landmarks and compare them with the database using data mining and use the Nearest Neighbor Approach for identifying similar images.

The chapter ends with deliberations on the future extent of Biometric technologies and the need to put in ample safeguards for data protection and privacy.

Keywords: Biometric, feature extraction, facial recognition, nearest neighbor approach

Corresponding author: siroyanavani@gmail.com

Pramod Singh Rathore, Vishal Dutt, Rashmi Agrawal, Satya Murthy Sasubilli, and Srinivasa Rao Swarna (eds.) Deep Learning Approaches to Cloud Security, (1–14) © 2022 Scrivener Publishing LLC

1.1   Introduction

Biometric authentication is a security process that relies on the unique biological characteristics of a person in order to affirm their identity. Biometric verification frameworks compare biometric data with existing original datasets that are stored. Examples of biometric characteristics include iris, palm print, retina, fingerprint, face, and voice signature. In recent years, deep learning-based models have helped accomplish best in class results in machine vision, audio recognition, and natural language processing tasks. These models appear to be a characteristic fit for dealing with the ever-expanding size of biometric acknowledgment issues, from phone verification to air terminal security frameworks. Thus, application of machine learning techniques to biometric security arrangements has become a trend [1].

Classification of Biometric Data:

- Behavioral Biometrics: gestures, vocal recognition, handwritten texts, walking patterns, etc.
- Physical Biometrics: fingerprints, iris, vein, facial recognition, DNA, etc.

Data science consultants can use machine learning's capacity to mine, look, and examine huge datasets for improving the execution of security frameworks and their reliability.

In light of its exceptional capacity to recognize people, biometric innovation has quickly become a way to help forestall shams and discovered its place in today's standard advancements. Consequently, it turns out to be more reliable than the customary validation frameworks that utilize passwords and documents for verification shown in Figure 1.1.

Physical modalities like fingerprints, voice, faces, veins, iris, hand geometry, and tongue print are unique and provide robust advancements in the field of cyber security [2]. They are useful compared to names, ID numbers, passwords, etc. because they are extraordinary, hard to reproduce, and are more significantly and genuinely bound to the individual.

A computing model which gives on-demand services like information stockpiling, computer power, and infrastructure to associations in the IT industry is termed to be "cloud computing" [3]. Despite the fact that cloud offers a ton of advantages, it slacks in giving security which is an issue for most clients. Cloud clients are hesitant to put classified information up because of looming threats to security.

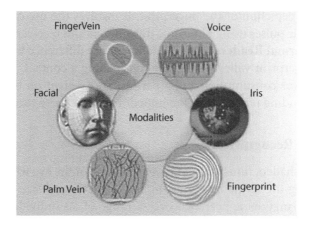

Figure 1.1 Biometric modalities [2].

1.2   Techniques of Biometric Identification

1.2.1   Fingerprint Identification

An automated technique for recognizing or affirming the identity of an individual dependent on the examination of two fingerprints is termed as Fingerprint Recognition. Human fingerprints are not easy to manipulate and are nearly unique and durable over a person's lifetime. They are unique, permanent, easy to acquire, and are a universally acceptable mode of identification [4].

Human fingerprints are difficult to control but remain sturdy over the life of an individual, making them suitable as long stretch markers of human character.

WORKING OF DIFFERENT TYPES OF FINGERPRINT READERS

1. **Optical Readers'** sensors work using a 2D image of the fingerprint. Algorithms can be utilized to discover novel patterns of lines and edges spread across lighter and hazier zones of the picture
2. **Capacitive Readers** use electrical signals to form the image of fingerprints. As the charges differ in the air gap between the ridges and lines in the finger set over the capacitive plate, it causes a difference in the fingerprint patterns.
3. **Ultrasound Readers** use high frequency sound waves to infiltrate the external layer of the skin which is used to capture a

3D depiction of the fingerprint. It involves the use of ultra-sonic pulses using ultrasonic transmitters and receivers.

4. **Thermal Readers** sense the temperature difference between fingerprint valleys and ridges on making a contact. Higher power consumption and a performance reliant on the surrounding temperature are impediments for these readers.

1.2.2   Iris Recognition

The iris is a shaded, flimsy, roundabout structure of the eye which controls light entering the retina by regulating the diameter and size of the pupil. It doesn't change its appearance over a range of an individual's lifetime except if harmed by external components [5]. Hereditarily indistinguishable twins also have distinctive iris designs. The irises of two eyes of an individual are also unique. **Iris recognition** is an automated method of identifying unique intricate patterns of an individual's iris using mathematical pattern-recognition techniques.

WORKING OF IRIS READERS

1. Scan an individual's eyes with subtle infrared illumination to obtain detailed patterns of iris.
2. Isolate iris pattern from the rest of the picture, analyze, and put in a system of coordinates.
3. Coordinates are removed using computerized data and in this way an iris mark is produced.

Even on disclosure, one cannot restore or reproduce such encrypted iris signatures. Now the user just needs to look at the infrared camera for verification. Iris acknowledgment results in faster coordination and is extremely resistant to false matches.

1.2.3   Facial Recognition

A non-intrusive technique to capture physical traits without contact and cooperation from people discovers its application in the acknowledgment framework. Every face can be illustrated as a linear combination of singular vectors of sets of faces. Thus, Principal Component Analysis (PCA) can be used for its implementation. The Eigen Face Approach in PCA can be utilized as it limits the dimensionality of a data set, consequently upgrading computational productivity [6].

WORKING OF FACIAL RECOGNITION TECHNIQUE:

Facial recognition technology identifies up to 80 factors on a human face to identify unique features. These factors are endpoints that can measure variables of a person's face, such as the length or width of the nose, the distance between the eyes, the depth of the eye sockets and the shape and size of the mouth. In order to measure such detailed factors, complexities such as aging faces arise. To solve this, computers have learned to look closely at the features that remain relatively unchanged no matter how old we get. The framework works by capturing information for nodal points on a computerized picture of a person's face and storing the subsequent information as a face print [7]. Face print is like a fingerprint but for your face. It accurately identifies the minute differences even in identical twins. It creates 3D models of your face and analyses data from different angles, overcoming many complexities associated with facial recognition technology. The face print is then utilized as a reason for correlation with information captured from faces in a picture or video.

1.2.4 Voice Recognition

Voice Recognition is a mechanized technique for recognizing or affirming the identity of an individual on the basis of voice. Voice Biometrics make a voiceprint for every individual, which is a numerical representation of the vocal tract of a speaker [8]. This is to ensure correct identification regardless of the language spoken, contents of speech, and wellbeing of an individual.

WORKINGS OF VOICE RECOGNITION:

1. Create a voice print or "template" of a person's speech.
2. Only when a user opts in or enlists himself, a template is created, encrypted, and stored for future voice verification.
3. Ordinarily, the enlistment process is passive, which means a template can be created in the background during a client's ordinary cooperation with an application or operator.

The utilization of voice biometrics for identification is expanding in fame because of enhancements in precision, energized to a great extent by evolution of AI, and heightened customer expectations for easy and fast access to information [9].

1.3   Approaches

Large amounts of data can diminish the efficiency of data mining and may not provide significant inputs to the model. Non-essential attributes add noise to the data, expanding the size of the model. Moreover, model building and scoring leads to consumption of time and system resources, influencing model precision.

Likewise, huge data sets may contain groups of attributes that are associated and may quantify a similar hidden component which can skew the logic of the algorithm. The computation cost associated with algorithmic processing increases with higher dimensionality of processing space, posing challenges for data mining algorithms. Impacts of noise, correlation, and high dimensionality can be minimized by dimension reduction using feature selection and feature extraction [10].

1.3.1   Feature Selection

Feature Selection selects the most relevant attributes by identifying the most pertinent characteristics and eliminating redundant information. A small size feature vector is used to reduce computational complexity, basic for online individual acknowledgment. Determination of effective features also helps increase precision [11]. Traditionally, large dimensions of feature vectors can be reduced using Principal Component Analysis (PCA) and Linear Discriminant Analysis (LDA).

The importance of a feature set S, for class c, is characterized by the normal estimation of all common data values between the individual feature f_i and class c as follows:

$$D(S,c) = \frac{1}{|S|} \Sigma_{f_i \in S} I(f_i; c) \cdot$$

The repetition of all features in set S is the normal estimation of all common data values between feature f_i and feature f_j:

$$R(S) \frac{1}{|S|^2} \sum_{f_i f_j \in S} I(f_i; f_j)$$

1.3.2   Feature Extraction

Feature extraction extracts distinct features from samples represented in a feature vector. Thus, alteration of attributes takes place in this method, whereas predictive significance criteria is used to rank the current attributes in feature selection technique. The altered features or attributes are linear aggregations of the original attributes. Since data is described by a lesser number of meaningful features, we obtain a higher quality model based on such derived attributes [12].

Feature extraction helps in data visualization by reducing a complex data set to 2 or 3 dimensions. It can improve the speed and efficiency of supervised learning. Feature extraction can also be used to enhance the speed and effectiveness of supervised learning. It has applications in data compression, data decomposition and projection, latent semantic analysis, and pattern recognition.

The information is extended onto the largest Eigen Vectors in order to reduce the dimensionality.

Let V = matrix with columns having the largest Eigen Vectors and D = original data with columns of different observations. Then, the projected data D' is derived as $D' = V^T D$.

In the event when just N Eigen Vectors are kept and $e_1...e_N$ represents the related Eigen Values, the amount of variance left after projecting the original d-dimensional data can be determined as:

$$s = \frac{\sum_{i=0}^{N} e_i}{\sum_{j=0}^{d} e_j}$$

1.3.3 Face Marking

Facial land marking and facial feature detection are significant activities that influence ensuing assignments concentrated on the face, for example, gaze detection, coding, face acknowledgment, and demeanor, as well as understanding of gestures, face tracking, and so on and so forth.

Face landmarks like the nose tip, eye corners, end points of the eyebrow curves, jaw line, nostril corners, and ear projections can serve as anchor points on a face chart. A few Landmarks that are less influenced by expressions are more reliable can be termed as fiducially points. In imaging systems, fiducially points are treated as imprints intentionally positioned in the scene to function as a point of reference shown in Figure 1.2 [13].

Applications of Face-Land Marking

- **Expression Understanding:** Facial expressions can be analyzed by means of temporal dynamics and spatial arrangements of landmarks.

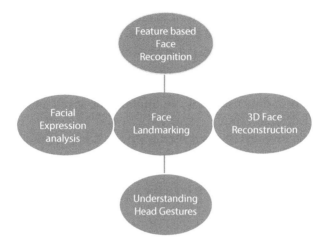

Figure 1.2 Applications of Face-Land marking [4].

- **Face Tracking:** Number of facial factors are depicted on the face graph model. Face tracking is acknowledged by letting the model chart advance as per shape of face parameters, facial segments, and their mathematical relations.
- **Face Recognition:** Locating the region of the eye and extracting holistic features from the windows fixated on different focal points.

1.3.4 Nearest Neighbor Approach

KNN Algorithm: A non-parametric regression and classification algorithm based on the model structure generated from data without any assumptions of its own.

KNN is used for measuring similarities by vector representation and comparison using an acceptable distance metric in various domains of data processing, pattern recognition, and intrusion detection. KNN is called memory-based or lazy learning in light of the fact that the manner in which it learns is simply storing representations of the training examples. An object is classified depending on the majority votes of its neighbors in the training set. The new model item will be ascribed to the class with its most comparable K-Closest Neighbors.

For facial acknowledgment, we can select the face descriptors and use the K-Nearest Neighbors (KNN) calculation to train our classifier.

Euclidean Distance Function of K-Nearest Neighbor can be used for feature extraction:

$$\sqrt{\sum_{i=1}^{k}(x_i - y_i)^2}$$

1.4 Related Work, A Review

P M Rubesh Anand (2018), along with other researchers, in his study, have done a complete analysis of the cloud environment and summed up the prevailing security threats in the cloud, conceivable outcomes, and alleviation in cloud administration with accentuation on access supervision, identity management, and services. Their research evaluates various facets with their commonly used techniques or mechanisms.

A K Jain (2012), in his research, sheds some light on the vulnerabilities of the biometric systems, the intrinsic limitations of the similarity of the any two biometrics, and its adversary effects. Unlike traditional authentication systems based on passwords, biometric authentications does not fully guarantee security.

A Patil (2018) in his paper discusses the security concerns of the cloud computing. Since cloud computing is a technology that delivers real time services, it is vulnerable to various kinds of data breaches. They discuss end to end communication through encryption, which would safeguard stolen information as the content would be encrypted and require security credentials [14].

To minimize, and ultimately beat, the dangers incorporated in the usage of customary strategies of authentication using PINs and passwords, a biometric framework for validation claimed to be more effective in controlling data breaches in cloud computing. In their research, they utilized an AES (Advanced Encryption Standard) algorithm for encrypting the data received from users at the time of enrollment and have devised another algorithm for the correlation of the user information with that of the layouts in the information database during authentication.

C S Vorugunti (2014) in his paper provides a simple and secure authentication system based on the SAAS model. It involves enrollment and verification as two steps of authentication. In the enrollment process, the biometric data is converted into a binary form. The feature extractor then converts the binary string into a set of features. In the verification process, the same feature will be processed when the user logins to the cloud. The process then verifies the cryptographic encryption and decryption operation on the users' biometric data.

S. Ziyad (2014) proposes in a study an authentication that is secured by the amalgamation of biometrics and cryptography. Their system structure involves three phases, namely, the initialization phase, registration phase, and verification phase. At the time of registration, biometric data is obtained from the users, encrypted, and stored in a smart card. Each smart card contains an authentication number along with palm vein biometric data and other related information. During the verification phase, the data from the smart card is verified with template data in the database and if the data is verified and matched, then the connection with the server is established and the user can access the system [15].

Traditional systems apply the authentication process in one or various modes. A single-sign-on is a strategy that utilizes customary techniques for the user to access the system just a single time upon entering their identity, however, they can access different services at different levels. S. Bawaskar (2016), in her research paper, proposes an upgraded SSO based authentication framework based on a multi-factor concept. The authors suggest a continuous bit sequence of the oriented certificates utilizing greater management schemes. Accordingly, the framework is totally secured, taking into account the need of protection from malicious activities.

1.5   Proposed Work

Facial acknowledgment is a classification of biometric programming that maps a person's facial features mathematically and stores the information as a face print.

Here, we propose a model for application in criminal justice systems. The model suggests the capacity to perform face identification in a group continuously or post-occasion, for open security, in urban communities, air terminals, at fringes, or other sensitive spots like religious congregations. It can help law enforcement agencies in better identification of possible suspects.

First of all, facial detection technique is used to confirm whether the image given as input is a face or not. On being detected as a face, facial marking is done using fiducially point analysis.

We will access the database from the cloud and the personal recognition will use the feature vectors from the training images to train or learn the classification algorithm. Data mining algorithms will be used for further processing and evaluation of data. Feature selection and feature extraction techniques can be used for improved accuracy shown in Figure 1.3.

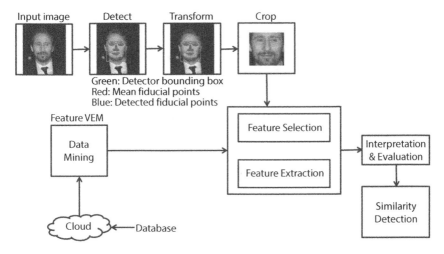

Figure 1.3 System architecture.

In the end, similarity detection is achieved by using the Nearest Neighbor Approach where the classifiers can be trained using a suitable algorithm like K-Nearest Neighbor.

STEPS:

1. An image captured is used as input.
2. Detect the image using fiducially points.
3. Transform the image by cropping the subsequent markers.
4. Apply Feature Selection and Feature Extraction for improved accuracy.
5. Use the Feature Vector Extraction method requiring features offloaded from the criminal database on the cloud.
6. Interpret and evaluate by comparing the two data sets.
7. Use the Nearest Neighbor Approach to identify the similarities between two or more images.

Facial Recognition Technology is intended to operate at a distance, without the knowledge of the target, so that it becomes hard to prevent the face from being captured. Along these lines, it allows for targeting of multiple persons one after another. Moreover, it is a non-consensual and clandestine reconnaissance innovation. The proposed model, when brought to realization, could act as an effective tool for criminal identification by comparing a live capture or digital image to the stored face print in order to confirm an individual's identity.

1.6   Future Scope

Biometrics pose danger to individual rights and privacy since technologies like facial recognition allow identification of citizens without their acknowledgement. Moreover, when consent is backed into the design of the technology, the privacy concerns regarding biometrics could be addressed [16].

Any modern technology is laden with concealed threats with no claim of infallibility either by the software maker, person selling it, or the one who advocates its deployment. In the context of criminal justice administration research, it indicates that images captured with default camera settings preferably expose fair complexion rather than dark, affecting results of Facial Recognition Technology across racial groups. One methodology might be to utilize a technology-neutral regulatory framework that identifies degrees of damages.

1.7   Conclusion

Biometric technologies have wide-ranging applications. They are being increasingly used every day for phone security, banks, and governments looking towards these technologies as security measures for verifying transactions. Important government organizations are using facial recognition technology to create databases using driver's license and passport details for effective administration, socio-economic development, and law enforcement. The thought that refined innovation implies more prominent proficiency should be fundamentally dissected. As these technologies penetrate more and more into our everyday lives, it is imperative to know and be educated about them. A reasonable strategy with ample safeguards for data protection and privacy is the need of the hour.

References

1. A.K. Jain and K. Nandakumar, "Biometric Authentication: System Security and User Privacy, "IEEE Published by the IEEE Computer Society, 0018-9162/12, Nov 2012.

2. Abhishek Kumar & Jyotir Moy Chatterjee & Pramod Singh Rathore, 2020. "Smartphone Confrontational Applications and Security Issues," International Journal of Risk and Contingency Management (IJRCM), IGI Global, vol. 9(2), pages 1-18, April.

3. Bhargava, N., Bhargava, R., Rathore, P. S., & Kumar, A. (2020). Texture Recognition Using Gabor Filter for Extracting Feature Vectors With the Regression Mining Algorithm. International Journal of Risk and Contingency Management (IJRCM), 9(3), 31-44. doi:10.4018/IJRCM.2020070103.

4. Chandra Shekharv Vorugunti, "A Secure and efficient Biometric Authentication as a service for cloud computing," 1 IEEE, October 09-11, 2014.

5. G.R Mettu and A. Patil, "Data Breaches as top Security concerns in Cloud Computing", International Journal of Pure and Applied Mathematics, 119(14):19-27,2018.

6. Haryatibinti Jaafar, Nordianabinti, Mukahar, Dzati Athiarbinti Ramli "A Methodology of Nearest Neighbor: Design and Comparison of Biometric Image Database" IEEE Student Conference on Research and Development, 2016.

7. Indu, P.M Rubesh Anand and V.Bhaskar, "Identity and Access management in Cloud Environment :Mechanisms and Challenges," Elsevier Engineering Science and Technology, an International Journal 21 (2018) 574–588, 2018.

8. K Sarat Chand and Dr. B Kezia. Rani, "Biometric Authentication using SaaS in Cloud Computing," International Research Journal of Engineering and Technology (IRJET), Volume: 05 Issue: 02, Feb-2018.

9. Kumar, A., Chatterjee, J. M., & Díaz, V. G. (2020). A novel hybrid approach of svm combined with nlp and probabilistic neural network for email phishing. International Journal of Electrical and Computer Engineering, 10(1), 486

10. N. Bhargava, S. Dayma, A. Kumar and P. Singh, "An approach for classification using simple CART algorithm in WEKA," 2017 11th International Conference on Intelligent Systems and Control (ISCO), Coimbatore, 2017, pp. 212-216, doi: 10.1109/ISCO.2017.7855983.

11. Naveen Kumar, Prakarti Triwedi, Pramod Singh Rathore, "An Adaptive Approach for image adaptive watermarking using Elliptical curve cryptography (ECC)", First International Conference on Information Technology and Knowledge Management pp. 89–92, ISSN 2300-5963 ACSIS, Vol. 14 DOI: 10.15439/2018KM19.

12. Rathore, P.S., Chatterjee, J.M., Kumar, A. et al. Energy-efficient cluster head selection through relay approach for WSN. J Supercomput (2021). https://doi.org/10.1007/s11227- 020-03593-4

13. S. Bawaskar and M. Verma, "Enhanced SSO based MultiFactor Authentication for Web Security", International Journal of Computer Science and Information Technologies, Vol. 7 (2), 2016, 960-966, 2016.

14. S. Ziyad and A. Kannammal, "A Multifactor Biometric Authentication for the Cloud", Springer India, Computational Intelligence, Cyber Security and

Computational Models, Advances in Intelligent Systems and Computing 246, 2014.

15. Singh Rathore, P., Kumar, A., & Gracia-Diaz, V. (2020). A Holistic Methodology for Improved RFID Network Lifetime by Advanced Cluster Head Selection using Dragonfly Algorithm. International Journal Of Interactive Multimedia And Artificial Intelligence, 6 (Regular Issue), 8. http://doi.org/10.9781/ijimai. 2020.05.003.

16. Thangapandiyan, M., Anand, P.M., & Sankaran, K. (2018). Quantum Key Distribution and Cryptography Mechanisms for Cloud Data Security. 2018 International Conference on Communication and Signal Processing (ICCSP), 1031-1035.

17. Joshua C. Klontz, Brendan F. Klare, Scott Klum, Anil K. Jain, Mark J. Burge, "Open source biometric recognition", Biometrics: Theory Applications and Systems (BTAS) 2013 IEEE Sixth International Conference on, pp. 1-8, 2012.

18. Patil, Archana and Patil, Dr. Rekha, An Analysis Report on Green Cloud Computing Current Trends and Future Research Challenges (March 19, 2019). Proceedings of International Conference on Sustainable Computing in Science, Technology and Management (SUSCOM), Amity University Rajasthan, Jaipur - India, February 26-28, 2019, Available at SSRN: https://ssrn.com/ abstract=3355151 or http://dx.doi.org/10.2139/ssrn.3355151

19. Sarvabhatla, M., Giri, M., Vorugunti, C.S., Cryptanalysis of cryptanalysis and improvement of Yan et al., Biometric- based authentication scheme for TMIS, CoRR, 2014.

20. Ziyad, S., & Rehman, S. Critical Review of Authentication Mechanisms in Cloud Computing, (2014).

21. Shruti Bawaskar et al, /(IJCSIT) International Journal of Computer Science and Information Technologies, Vol. 7 (2), 960-966, 2016.

Privacy in Multi-Tenancy Cloud Using Deep Learning

Shweta Solanki[1*] and Prafull Narooka[2]

[1]MDS University Ajmer, Ajmer, India
[2]Department of Computer Science, Agrawal College, Merta City, Rajasthan

Abstract
There is a responsibility to maintain the privacy and security of data in the Cloud Computing environment. In present times, the need for privacy is increased due to frequent development in multi-tenant service based systems. As a system of growth increases, the requirement for privacy also increases. We use Deep Learning concepts to increase privacy levels. In this chapter, we understand the cloud computing concept within a Multi-Tenant Framework (MWF). In Multi-Tenant Frameworks, requirements for privacy and security concepts are developed using Deep Learning. The goal is to find privacy requirements across many factors in a Multi-Tenancy based systems using Deep Learning concepts. The services of Multi-Tenant based systems are aggregated due to the dynamic environment of Cloud Computing. Three consistencies will be maintained by privacy policies using Deep Learning. In Multi- Tenancy, a large number of users (tenants) use the same services required for privacy and security to maintain the durability and consistency of service.

Keywords: Multi-tenant, privacy, framework, privacy policy, cloud computing, single tenant, public, private

2.1 Introduction

It is very important to understand the need for Multi-Tenancy in the Cloud Computing environment because today all work is done with the help of an

---

Corresponding author: shweta.solanki01212@gmail.com

Pramod Singh Rathore, Vishal Dutt, Rashmi Agrawal, Satya Murthy Sasubilli, and Srinivasa Rao Swarna (eds.) Deep Learning Approaches to Cloud Security, (15–26) © 2022 Scrivener Publishing LLC

internet transaction mode. In this environment, the database and schema used in the database change or modify frequently. Data is stored in different areas in different databases [1]. The cloud computing environment of every organisation is different. The real need of cloud environment-like private or public modes depends on the concept of data and tenant needs, organisation needs, or the configuration of a database depending on the structure of the database and the model used by the organisation for privacy policies. The security level of the Multi-Tenancy structure should maintain the durability of an application to maintain consistency of the system and avoid interruption of the regular based work done by the tenant. The concept of Multi-Tenancy in multiple databases requires more privacy between tenants both logically and physically to accommodate a greater the need for privacy and security in each work area. Using Deep Learning concepts can improve the complexity in accessing and managing a database. Using Deep Learning concepts should reduce access and increase efficiency, preventing the leak of data, hacking, or other risks because they can be found out easily and improve the quality of Multi-Tenant system efficiency. So, in Deep Learning, the privacy concept is improved and the transparency between the multiple tenants increases, maintaining privacy. Deep Learning concepts are very powerful and scalable for implementing. In Deep Learning, databases are managed efficiently. The Deep Learning algorithm makes this affordable. The Deep Learning concept has the ability to improve data driven predictions. It finds the patterns of privacy and security in databases in Multi-Tenancy. This can provide good or better values for organisation. In Deep Learning, the work is done for binary, category, and value predictions in Cloud Computing. In this chapter, we discuss the basic concept of Multi-Tenancy, privacy requirements, and the Cloud Computing concept with Deep Learning.

2.2   Basic Structure

It is very important to understand Cloud Computing, as it provides service accordingly to its user end. Whether it is a private or public cloud depends the requirement of user, like Multi-Tenancy or Single Tenancy. It is also an on demand service that depends on tenant requirement, resource availability, storage requirement, activity management, and which topology is required for the distributed system. The tenant requires either a centralised or decentralised framework. Security in Cloud Computing also depends on the concept of the database and the need for bigness and organisation. As the need arises, the organisation selects the structure of Cloud Computing.

Then, they create the schema and select a data model and select a Single or Multi-Tenancy concept for work. Complexity and cost are also dependent on the requirements of the organisation's needs. There are structures of Cloud Computing and the Multi-Tenancy concept available illustrating the impact of Multi-Tenancy in Cloud Computing. The cloud manages the concept of Multi-Tenancy, shares resources, and manages services with many tenants. Management and the utility services are provided by the structure according to need. This section discusses the basic structure of Cloud Computing and how the work is done in combination with Multi-Tenancy [2].

2.2.1 Basic Structure of Cloud Computing

Cloud Computing services are used by every service area, whether the business is small or large.

The cloud facility is available according to the requirements of the business. The cloud enables the facility to store data, whether it be large or small, and provides access to data from any location and any hardware or a virtual environments for accessing or storing data.

Cloud Computing provides the services of both service oriented and event driven architecture.

The given diagram represents the various services of the cloud. All services are inter-related to each other. The three main parties involved are the cloud service consumer, cloud service provider, and cloud broker shown in Figure 2.1.

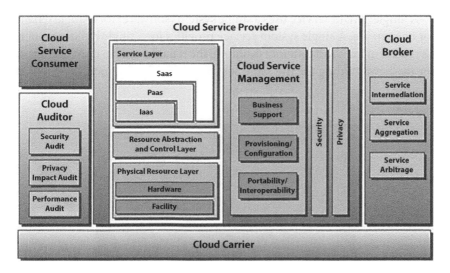

Figure 2.1 Cloud computing services [3].

In cloud service, all consumer services are provided if the consumer belongs internally or, if the consumer belongs externally, it depends on the consumer requirement. At the time of consumer requirement, the service provides, as the diagram represents, the work of a cloud auditor. The cloud service provider provides all cloud services. Whether the type of service provided by the service provider is commercial or corporate is decided by the service provider [3].

The cloud service provider can provide all services with physical or virtual resources that provide the cloud service and create applications according the requirement of services. All groups work together as a service orchestration.

2.2.2   Concept of Multi-Tenancy

The basic concept of Multi-Tenancy is required when many tenants work together with the use of internet and data can be transferred using digital services if the tenants are in different places; the concept is to work from one place to another place. This service totally depends on the needs of the organisation. The requirement of tenants and which type or place the tenant requires is up to the organisation's needs [4].

Basically, there are two types of tenancy in an SaaS environment: Single and Multi-Tenancy. Figure 2.2 briefly discusses Multi-Tenancy's use in real time application. In Single Tenancy, a single environment or platform is used, but in Multi-Tenancy, a common platform is used in different places or shares the platform of work and creates a known work place for a single organisation.

A Single Tenant service uses a single system (software) for a single tenant for one service. A single system service provides for a single building, but if its requirement is in a different building, the tenant needs to purchase again for the new building for the same services. In Multi-Tenancy, the client is

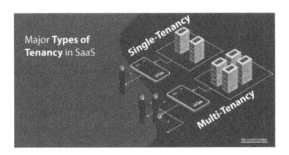

Figure 2.2 Types of tenant [4].

not required to purchase a single service multiple times, which means single services can be used in different offices of a single organisation.

In a Multi-Tenancy system, it is easy to modify and make changes as required by the tenant. Multi-Tenant systems combine the work place between the client and the tenant, but the concept or logic should be separate to each other. The organisation shares the data and configurations, as well as services of management with the user and tenant [5].

There are basically three major concepts in Multi-Tenancy, as show in the Figure 2.2.

1. A **shared database** which is used to manage the Multi-Tenancy system means this facility helps make data available and manages the cost or storage area flexibility and scalability management. It reduces the complexity of managing the database and makes it easily available to each tenant and workspace.

2. **One database and many schemas** means all tenants use the same database in a different manner. Each tenant uses their requirement of data and every tenant requirement is different as the work is different. According the tenant, the services are provided to the tenant. The complexity and cost will also affect the structure used by the tenant.

3. **Many databases and schemas** means, in this type of model, that tenant data is stored in the database in different locations in a different database or that the tenant can create different databases as they are required. Accordingly, the new database will create a cost and complexity increase [6].

2.2.3   Concept of Multi-Tenancy with Cloud Computing

In the Cloud Computing system, the concept of Multi-Tenancy is very important because IT provides the facility to all tenants through shared computer resources in a cloud environment. The cloud environment has two types: a public cloud environment and a private cloud environment. The tenant chooses one of the cloud environments. All tenant data is isolated and inaccessible to each other shown in Figure 2.3.

Multi-Tenancy cloud computing systems create a sparse work area for each tenant for storage or project data and login privacy policies. The tenant only uses their personal and secure area for work and accesses only know work or data. In the case of other requirements of data, another tenant's permission or access key is required for access.

One type of cloud computing is in the private cloud environment and uses groups of more than two tenants belonging to a single company.

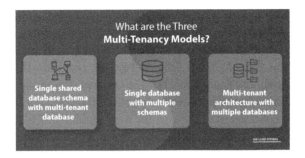

Figure 2.3 Multi-Tenancy models [5].

They work on the same data and use resources through the cloud environment. The other type is the public cloud environment where many companies can share their data and services with different tenants. The public cloud environment is used more than the private cloud environment.

The Single Tenant Cloud Computing environment does not provide to all facilities. When compared to a Multi-Tenancy system, it provides more storage, access features, and security and privacy policies.

It provides a virtual environment for work easily, with less complexity maintaining work and hardware and no restrictions to devices or location [7].

Example of Multi-Tenancy with Cloud Computing:
Let us take a very common example of a colony. In a colony there are many houses; each house has many flats, each house is playing a role as part of a single company, each company has different departments, each

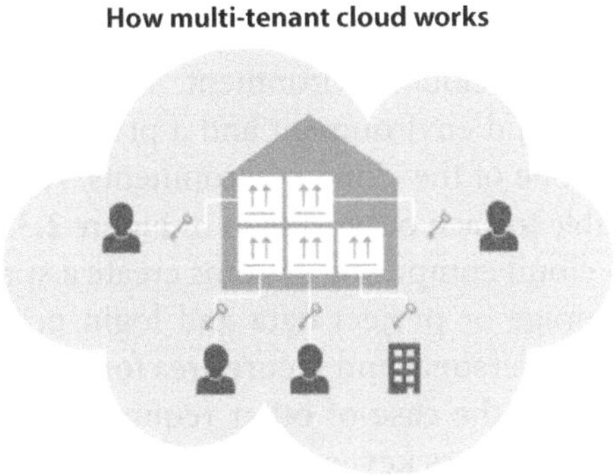

How multi-tenant cloud works

Figure 2.4 Multi-Tenant Cloud structure [6].

department works only at their area, and each department shares common utility services (i.e. water, electricity, PNG etc.). Similar facilities are used by companies shown in Figure 2.4.

2.3 Privacy in Cloud Environment Using Deep Learning

In Multi-Tenant cloud based systems, security uses the Deep Learning Method to overcome all the requirements of a tenant, including privacy policies and services required. Using Deep Learning Methods for developing the privacy structure of Cloud Computing provides better services according to the requirements of a tenant. Accordingly, at the request of the tenant, resources and service availability fulfil requests and the privacy and security services are developed and maintained. In the cloud computing environment, there are public, private, single, and multi-tenant structures available to use according to the requirements of a tenant. Different structures have different needs for privacy and security services. If better services are not available, the tenant may not use the structure, therefore, using Deep Learning methods develops privacy and security services. The services are developed according to the organisation's needs using Deep Learning [8].

Privacy in Cloud Computing is maintained by a distributed system concept in a Multi-Tenancy based application using encryption techniques to protect data, synchronize work, and regulate base data modification if the tenant is authorized or not, so all factors are required to check and manage [9]. The privacy management system is used to manage the security of sophisticated data access.

The information stored in the cloud environment is directly encrypted in a format that the tenant is authorised to access and the data can only decrypt that data securely [10]. Using Deep Learning techniques increases privacy and security levels. A new protection algorithm has been developed using Deep Learning Methods for safe transformation and using data in a database in a cloud environment. In a Cloud Computing environment, a privacy service system has been developed to stop unauthorized access and increase the capabilities for the security of private data, minimizing the hacking of private data in the cloud environment shown in Figure 2.5 and Figure 2.6 [11].

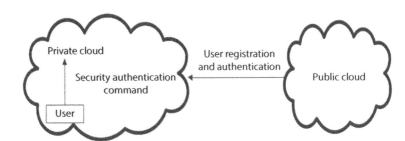

Figure 2.5 Concept of Cloud Security [7].

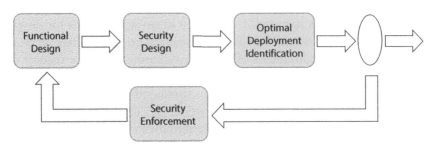

Figure 2.6 Securities in Multi-Cloud environments [7].

2.4 Privacy in Multi-Tenancy with Deep Learning Concept

There is a need for privacy in a Multi-Tenancy system because of the risk of low privacy policies and weak security for data when tenants work on multi-tenant applications. The organisation and tenant are not able to self-secure all data [12]. Using Deep Learning creates a concept to secure data, providing there is a secure and private environment for the tenant to accept the Multi-Tenant application and work on that system freely. This is the reason for weak privacy and security polices, a risk of data loss, a risk of hacking of data, and the wrong use of information, so it is very necessary to secure the entire task before starting work on a Multi-Tenant system. For security or privacy, the first step is maintaining the concept of a unique ID [13]. In this model, all tenants have an individual, unique ID for login. If the organisation is very large and it is complicated to manage all the IDs, then each individual department will have a single ID to login. This concept is also used for using Deep Learning. The second step is to make access limitations for each tenant. The access of each tenant is dependent on the organisation or authorised department deciding the access limitation. For example, a company whose tenants are working in an account department are able to access only the account

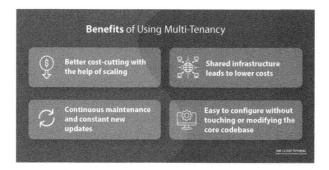

Figure 2.7 Multi-Tenancy services [8].

department data, they are not able access other departments' (admin, security, etc.) data. According to this concept, all department access criteria is decided or fixed and department tenant access limitation is decided so the authorisation is checked and only authorised users can access the limited data. The third step is to isolate the database into tables according to department. The database is then separated and isolated into tables according to tenant access and limitation-isolated data is provided to the tenant. From this method, the whole database is not given to all tenants and only isolated data is provided by using a Deep Learning concept shown in Figure 2.7.

The fourth step is encryption in a Multi-Tenant based system. In a Multi-Tenant based system, the consistency, integrity, durability, accuracy, and on-time demand of a database is mandatory for fulfilment. If the Multi-Tenant based system does not fulfil the requirements due to any term and condition, the tenant may not be able to work efficiently in the organisation, so encryption techniques based on Deep Learning concepts are used to secure the database. Some encryption techniques include digital, security, key, signature, digital key, private key, and password provided encryption [14].

The authorised user accesses the sophisticated database and can modify the database. If unauthorised access happens, sophisticated data access by the unauthorized user can be added, deleted, and modified by unauthorised activity. In a Multi-Tenant system using encryption techniques, first check the authorisation, find out if the tenant is authorised or not, and if the tenant has been authorised as a user with the access provided.

2.5 Related Work

In this chapter, we will look at the work related to the concept of Multi-Tenancy privacy policies. Future use of Multi-Tenancy in the cloud environment is dependent on the complexity and cost affected to the data structure. There are many works done in many chapter basics in privacy and security concepts [15] where database hacking and transition fraud happened. The use of Deep Learning removes that type of problem and reduces fraud. This chapter data is useful to find out the functional or non-functional parameters of clouding computing systems with respect to Multi-Tenant systems. These details take discretion from parameters like security and privacy concepts, detail descriptions on the structure of Multi-Tenancy in cloud based frameworks, vary modules of Multi-Tenancy use according to requirements, and discuss the security, privacy, performance, cost, and flexibility factors of Multi-Tenancy cloud based systems. This chapter also discusses the contributions of Deep Learning concepts used in data security and privacy and in protection concepts as cloud computing system architecture. This chapter is used to find the maximum solution to protect and maintain the privacy and security of databases and the work place of tenants in Multi-Tenancy based systems using Deep Learning concepts and understand the structure of cloud computing and deep structure of Multi-Tenancy with a privacy concept of Deep Learning methods. This literature is used to understand and find the requirements of resources, services, and privacy concept development in various services like response time, network load, and throughput management services development, as well as the need for resources and requirement of resources in a cloud based Multi-Tenant system and privacy services using Deep Learning concepts [16].

2.6   Conclusion

In Cloud Computing with a Multi-Tenancy system, privacy and security are very complicated and valuable. These concepts are important because it is a responsibility to provide privacy and security to the unique architecture of cloud computing and multi-tenant systems. The data must be correct, durable, and secure. Every tenant wants to work in a secure environment; this helps create a good and graceful environment for the work place. Every tenant wants security and privacy to be maintained for database transactions in the cloud environment. If this requirement is not fulfilled, the tenant will work for longer durations and the ability for work will reduce, so privacy and security are two factors which decide the future of that structure. Using the Deep Learning concept, we work on the privacy and security areas of Multi-Tenancy systems making them more secure for both physical and logical separation and also provide a great privacy platform to work free from any worry about security. With the help of Deep

Learning, the Multi-Tenant system makes things more secure and privacy policies more stable to work with and secures the future safety of the database used by different tenants of the same organisation. Using the Deep Learning concept provides mechanisms to make privacy architecture to enhance the security level of privacy policies. Data is secure on the front and back ends, so the isolation of data is protected at both ends and is safe for future use by the tenant. It is sophisticated and necessary for the privacy and security of each face of a cloud based Multi-Tenant system to maintain no loss of data for the durability and safe side of a system.

References

1. Abhishek Kumar & Jyotir Moy Chatterjee & Pramod Singh Rathore, 2020. "Smartphone Confrontational Applications and Security Issues," International Journal of Risk and Contingency Management (IJRCM), IGI Global, vol. 9(2), pages 1-18, April.
2. Bhargava, N., Bhargava, R., Rathore, P. S., & Kumar, A. (2020). Texture Recognition Using Gabor Filter for Extracting Feature Vectors With the Regression Mining Algorithm. International Journal of Risk and Contingency Management (IJRCM), 9(3), 31-44. doi:10.4018/IJRCM.2020070103
3. By Judith Hurwitz https://www.dummies.com/programming/cloud-computing/ hybrid-cloud/multi-tenancy-and-its-benefits-in-a-saas-cloud-computing-environment/ by Judith Hurwitz, Marcia Kaufman, Fern Halper, Daniel Kirsch
4. https://www.dummies.com/programming/cloud-computing/hybrid-cloud/ multi-tenancy-and-its-benefits-in-a-saas-cloud-computing-environment/
5. Computer term http://whatis.techtarget.com/definitionmulti-tenancy Frederick Chonghttps://msdn.microsoft.com/en-us/library/aa479086.aspx
6. https://www.researchgate.net/publication/311922746
7. Kumar, A., Chatterjee, J. M., & Díaz, V. G. (2020). A novel hybrid approach of svm combined with nlp and probabilistic neural network for email phishing. International Journal of Electrical and Computer Engineering, 10(1), 486
8. Margaret Rouse https://searchcloudcomputing.techtarget.com/definition/ multi-tenant-cloud.
9. Multi tenancy in SaaS-PaaS http://multitenancy-in-saaspaas.wikispaces.asu.edu/
10. Naveen Kumar, Prakarti Triwedi, Pramod Singh Rathore, "An Adaptive Approach for image adaptive watermarking using Elliptical curve cryptography (ECC)", First International Conference on Information Technology and Knowledge Management pp. 89–92, ISSN 2300-5963 ACSIS, Vol. 14 DOI: 10.15439/2018KM19.
11. Rathore, P.S., Chatterjee, J.M., Kumar, A. et al. Energy-efficient cluster head selection through relay approach for WSN. J Supercomputer (2021). https:// doi.org/10.1007/s11227-020-03593-4

12. Shildshare.net//www.slideshare.net/mmubashirkhan/saa-s-multitenant database- architecture
13. Singh Rathore, P., Kumar, A., & Gracia-Diaz, V. (2020). A Holistic Methodology for Improved RFID Network Lifetime by Advanced Cluster Head Selection using Dragonfly Algorithm. International Journal Of Interactive Multimedia And Artificial Intelligence, 6 (Regular Issue), http://doi.org/10.9781/ijimai.2020.05.003.
14. www.elsevier.com/locate/knosys
15. www.researchgate.net/publication/320185699
16. Thangapandiyan, M., Anand, P.M., & Sankaran, K. (2018). Quantum Key Distribution and Cryptography Mechanisms for Cloud Data Security. 2018 International Conference on Communication and Signal Processing (ICCSP), 1031-1035.

3

Emotional Classification Using EEG Signals and Facial Expression: A Survey

S J Savitha*, Dr. M Paulraj and K Saranya

Department of Computer Science and Engineering, Sri Ramakrishna Institute of Technology, Coimbatore, India

Abstract

Electroencephalogram (EEG) and facial expression based emotion classification is a relatively new field in the affective computing area with challenging issues regarding the consecration of emotional states and extraction of features in order to attain optimum classification performance. Automatic emotion detection using facial expression recognition is now a main area of various fields such as computer science, medicine, and psychology. Various feature extraction techniques have been developed for classification of expressions from EEG signals from brain and facial expressions from static images, as well as real time videos. Deep learning plays an important role for this kind of task. This chapter provides a review of research work carried out in the field of facial expression recognition and EEG signal classification.

Keywords: Electroencephalogram (EEG), brain computer interface (BCI), emotion recognition, feature optimization, evolutionary algorithm

3.1   Introduction

The possibility of establishing a direct communication and control channel between the human brain and computers or robots has been a topic of scientific speculation and even science fiction for many years. Over the past twenty years, this idea has been brought to fruition by numerous research

---

Corresponding author: savitha.cse@srit.org

Pramod Singh Rathore, Vishal Dutt, Rashmi Agrawal, Satya Murthy Sasubilli, and Srinivasa Rao Swarna (eds.) Deep Learning Approaches to Cloud Security, (27–42) © 2022 Scrivener Publishing LLC

and development programs and has evolved into one of the fastest-growing areas of scientific research. This technology, called brain-computer interface (BCI) technology, provides a new output channel for brain signals to communicate or control external devices without using the normal output pathways of peripheral nerves and muscles. EEG has been extensively employed for BCI applications due to its non-invasiveness, ease of implementation, and cost-efficiency. Recently, dry-electrodes and wireless EEG systems have also been studied in various experimental settings which has further expanded the scope of EEG based BCI systems. Deep learning methods provide many ways to accomplish this kind of task. As frequency analysis of EEG is a major concern, many approaches, such as the band pass filter bank, autoregressive (AR) model, Fourier transform, and Wavelet transform, have been employed. However, all these methods have certain limitations. The frequency resolution of a band-pass filter is constrained by the order of the filter. The AR model in deep learning coefficients only provide limited frequency information as the envelope of the spectrum reconstructed from AR coefficients is limited by the order of AR model. The discrete wavelet transform is considered a band-pass filter bank with orthogonal basis that has better frequency and temporal resolution. However, the obtained wavelet coefficients are redundant in general. Further, post-processing is required to stabilize the performance of the classifier. The EEG signal considered for BCI is generally band-limited. To enhance classification performance, more data from EEG channels is often required for more complex problems. The increase in the number of EEG channels poses two challenges. First, as the number of channels increases, direct identification of the subject-dependent reactive band becomes infeasible and an automated identification process is required. Secondly, as the feature vector includes all frequency estimates from all the channels, the dimension increases drastically. Deep Learning provides high dimensionality in feature vectors that will affect the performance of the classifier [1].

A BCI allows a person to communicate with or control the external world without using conventional neuromuscular pathways. That is, messages and control commands are delivered not by muscular contractions, but rather by brain signals themselves. This BCI feature brings hope to individuals who are suffering from the most severe motor disabilities, including people with amyotrophic lateral sclerosis (ALS), spinal cord injury, stroke, and other serious neuromuscular diseases or injuries. BCI technology holds promise to be particularly helpful to people who are "locked-in" cognitively but without useful muscle function. Restoration of basic communication capabilities for these people would significantly improve their quality of life as well as that of their care-givers, increase independence, reduce social isolation, and potentially reduce cost of care [2].

BCI research has undergone an explosive growth in recent years. At present, there are over 400 groups worldwide engaging in a wide spectrum of research and development programs using a variety of brain signals, sig-nal features, and analysis and translational algorithms. In this review, we discuss the current status and future prospects of BCI technology and its clinical applications. We will define BCI, review the BCI-relevant source signals from the human brain, and describe the functional components of BCIs. We will also review current clinical applications of BCI technology and identify potential users and potential applications of Deep Learning. Finally, we will discuss current limitations of BCI technology, impedi-ments to its widespread clinical use, and expectations for the future in the era of Deep Learning.

3.2   Related Works

There are several works existing related to human emotion analysis.

Evolutionary Computation Deep Learning Algorithms for Feature Selection of EEG Based Emotion Recognition Model. In this model, EEG features are extracted in three ways from time, frequency, and time fre-quency domains. There is no standardized set of features that have been generally agreed upon as the most suitable for emotion recognition. This leads to what is known as a high-dimensionality issue in EEG, as not all of these features carry significant information regarding emotions. Irrelevant and redundant features increase the feature space, making pattern detec-tion more difficult and increasing the risk of over fitting. It is therefore important to identify salient features that have significant impact on the performance of each emotion [3].

Adaptive Neuro-Fuzzy Interference Model in Deep Learning. The pro-posed ANFIS model combined neural network adaptive capabilities and the fuzzy logic qualitative approach to improve diagnostic accuracy. The WT can be applied to extract wavelet coefficients of discrete time signals. This procedure makes use of multi-rate signal processing techniques. The proposed scheme is a sub band coding or multi-resolution signal analysis. The multi-resolution feature of the WT allows decomposition of a signal into a number of scales, each scale representing a particular coarseness of the signal under study [4]. The WT provides very general techniques that can be applied to many tasks in signal processing. One very important application of the Deep Learning model is the ability to compute and manipulate data in compressed parameters.

Artificial Neural Networks (ANNs) in Deep Learning have been used as computational tools for pattern classification, including diagnosis of diseases, because of the belief that they have greater predictive power than signal analysis techniques. In this study, a new approach based on ANFIS was presented for the classification of EEG signals. The proposed technique involved training the five ANFIS classifiers to classify the five classes of EEG signals when wavelet coefficients defining the behavior of the EEG signals were used as inputs. The ANFIS classifiers were trained with backpropagation. Each of the ANFIS classifiers was trained so that they are likely to be more accurate for one class of EEG signals than the other classes. The predictions of the five ANFIS classifiers were combined by a sixth ANFIS classifier [5]. The correct classification rates and convergence rates of the proposed ANFIS model were examined and the performance of the ANFIS model of Deep Learning was better than other neural network methods.

The fractal dimension method for the analysis of physical and biological time domain signals has high complexity and irregularity. In this chapter, they used a fractal dimension of neural networks as a feature instead of a linear feature extraction technique of Deep Learning, for example calculation of a power spectrum. In this chapter, they used the frequency dimension to measure the complexity and irregularity of EEG signals, which is also a suitable method for detecting fast transients in EEG signals [6]. Calculating of FD is done directly from EEG time series and it does not require the reconstruction of the attractor in the state space of multidimensionality. In Deep Learning, there are many techniques available in literature to compute the FD. However, Katz and Higuchi algorithms are widely used and produce better results on EEG signals.

The different kind of emotions and their EEG features find the exact feature for emotions and identify the accuracy for emotions using different machine learning techniques like WT and SVM. Here, the emotions are grouped based on being enjoyable and disagreeable.

Basic emotions like happy and sad are used to extract other emotions, so the emotions can be identified using LDA and KNN. The features are extracted using discrete Wavelet Transformation and SAM. It deals with both symmetric and asymmetric levels of emotions. The emotions can be identified by using signal segmentation and extracting the feature using HOC and the Cross Correlation Method.

SVM in Deep Learning is used to classify emotions into different groups like negative and positive emotions. By using this method, we have to get the best accuracy of emotions and multidimensional data also used. Emotions are classified from different optimized techniques and those emotions are related to human machine interaction and other activities which are used in affective computing. The features are extracted from EEG using HOC. It supports different feature extraction techniques and provides better accuracy results. The emotions are identified using short term assessment and particular time related emotions, are only extracted using STFT and MI, and are classified using SVM and RVM [7].

All these techniques and methods are implemented using MATLAB and different types of people EEGs are used to analyze the emotions. The EEG signal has records of healthy persons without any psychotic disease shown in Table 3.1.

Table 3.1 Feature extraction techniques [8].

S. no.	Method	Contribution	Issues
1.	Wavelet Transform	1. Analyzes the signal with variable window 2. Analyzes both time and frequency data	1. Lacking specific method 2. Performs with limited Heisenberg uncertainty
2.	Principal Component Analysis	Analysis reduces dimensionality without loss of information	1. Data is complicated 2. PCA fails to process this data
3.	Non - Linear Dynamic Analysis	1. Computationally efficient for large data 2. Decomposes signal into temporal data	1. Requires more computation 2. Does not work with determinate cases
4.	Short Time Fourier Transformation & Mutual Information	Fixed slide window length	1. Cannot denoise 2. Tradeoff between time and frequency
5.	Higher Order Crossing	1. Performance should be high 2. Provides optimized result	1. Difficult to choose the random data

3.3   Methods

Deep Learning provides ways in which the EEG data are analyzed using several procedures, including signal preprocessing, feature extraction, dimensionality reduction, and classification methods to find emotions [8].

3.3.1   EEG Signal Pre-Processing

Pre-processing and feature extraction are two main steps in EEG signal processing. Pre-processing techniques help remove unnecessary artefacts from the EEG signal and increase the signal to noise ratio. A preprocessing block aids in improving the performance of the system by separating noise from the actual signal. Before processing the EEG, the signal has to remove noise and unwanted details by using spectral filtering and surface Laplace Transformation. 4 to 10Hz frequencies are used to extract the emotion related features. The following Feature Extraction Techniques of Deep Learning are used.

3.3.1.1   Discrete Fourier Transform (DFT)

In a time domain, representation of digital signals describes the signal amplitude versus the sampling time instant or the sample number. However, in some applications, signal frequency content is very useful otherwise as digital signal samples. The representation of the digital signal in terms of its frequency component in a frequency domain, that is, the signal spectrum, needs to be developed [9]. The algorithm transforming the time domain signal samples to the frequency domain components is known as the Discrete Fourier Transform, or DFT. The DFT also establishes a relationship between the time domain representation and the frequency domain representation. Therefore, we can apply the DFT to perform frequency analysis of a time domain sequence. In addition, the DFT is widely used in many other areas including spectral analysis, acoustics, imaging and video, audio, instrumentation, and communications systems.

3.3.1.2   Least Mean Square (LMS) Algorithm

Conventional filtering cannot be applied to eliminate artefacts because EEG signals and artefacts have overlapping spectra. An adaptive process was created in which the transfer function H(z) is adjusted according to an optimizing algorithm. The adaptation is directed by the error signal between the primary signal and the filter output. The most used optimizing criterion is the Least Mean Square Algorithm. This algorithm of Deep Learning is an application scheme widely used in practice due to its simplicity [10].

3.3.1.3   Discrete Cosine Transform (DCT)

The DCT is a transform that is very common when encoding video and audio tracks on computers. It has found applications in digital signal processing and particularly in transform coding systems for data compression and decompression. DCT takes correlated input data and concentrates its energy in just the first few transform coefficients [11].

3.3.2   Feature Extraction Techniques

Feature extraction is the process of identifying the particular information from an EEG, which was measured by brain activity. This chapter focuses on several feature techniques, including:

- Discrete Wavelet Transformation
- Higher Order Crossing
- Principal Component Analysis
- Independent Component Analysis
- Short Time Fourier Transform and Mutual Information
- Time and Frequency Attribute

This table explains different feature extraction methods and their advantages and disadvantages, which is helpful, for knowing how the feature is extracted from the EEG signal and analyzing emotion related features. By using this table, we can identify particular characteristics of the methods and the method that is extracting particular features with better results during the extraction process [12].

3.3.3   Classification Techniques

After extracting the features, classification is used to group related emotions. The following methods of Deep Learning are used for classification:

- Neural Network
- K-Nearest Neighbor
- Support Vector Machine
- Linear Dynamic Analysis

Table 3.2 Classification techniques.

S. no.	Method	Contribution	Issues
1.	Neural Network	1. Easy to train 2. Accurate pattern classification	1. Needs training to operate 2. Requires high processing time for large network
2.	K-Nearest Neighbor	1. Easy to understand 2. Easy to implement	1. Poor runtime performance 2. Sensitive to irrelevant and redundant features
3.	Support Vector Machine	1. Good generalization 2. More performance	1. Computational complexity is high
4.	Linear Dynamic Analysis	1. Extremely fast 2. Low requirements 3. Good results	1. Fail to discriminate functions like variety of features 2. Complex structure

This Table 3.2 explains the different feature classification methods and their advantages and disadvantages, which is helps in knowing how the features are extracted from the EEG signal and analyzing emotion related classes [13].

3.4 BCI Applications

Individuals who are severely disabled by disorders such as ALS, cerebral palsy, brainstem stroke, spinal cord injuries, muscular dystrophies, or chronic peripheral neuropathies might benefit from BCIs. To help determine the value of BCIs for different individuals, potential BCI users are categorized by the extent, rather than the etiology, of their disability. Evaluated in this way, potential BCI users fall into three reasonably distinct groups: (1) people who have no detectable remaining useful neuromuscular control and are totally locked-in; (2) people who retain only a very limited capacity for neuromuscular control such as weak eye-movements or a slight muscle twitch; and (3) people who still retain substantial neuromuscular control and can readily use conventional muscle-based assistive communication technology [14].

It is not yet clear to what extent BCIs can serve people in the first group, those who are totally locked-in (e.g., by late-stage ALS or severe cerebral palsy). Resolution of this issue requires extensive and prolonged evaluation of each individual in order to resolve basic issues of alertness, attention, visual or auditory capacities, and higher cortical function. While it has been hypothesized that the totally locked-in state constitutes a unique BCI-resistant condition, the issue remains unresolved at present. It is worth mentioning that researchers have speculated that individuals in this group might be able to retain the capacity for BCI use if they begin it before becoming totally locked-in.

At present, people in the second group constitute the primary prospective user population for current BCI systems. This group, which outnumbers the first group, includes people with late-stage ALS, patients who rely on artificial ventilation as their disease progresses, people with brainstem strokes, and people with severe cerebral palsy. Typically, they retain only very limited, easily fatigued, and/or unreliable eye movements or other minimal muscle function and, thus, cannot be adequately served by conventional muscle-based assistive communication technology. For people in this group, BCI systems may be able to provide basic communication and control that is more convenient and reliable than that provided by conventional technology in the Deep Learning domain [15].

The third and largest group of potential BCI users consists of people who retain substantial neuromuscular control [16]. For most in this group, present-day BCI systems, with their limited capacities, have little to offer. These individuals are usually much better served by conventional technology. Nevertheless, some in this group, such as those with high-cervical spinal cord injuries, may prefer a BCI over conventional assistive devices that coopt their remaining voluntary muscle control (e.g., systems that depend on gaze direction or EMG from facial muscles). In the future, as the capacities, reliability, and convenience of BCI systems continue to improve, more people in this group could find them of value and the number of people using BCIs could substantially increase [17].

The different conditions mentioned above impair the CNS in different ways and different BCIs depend on different aspects of brain activity. Thus, some people may be better served by one BCI than by another. For example, people who have sensorimotor cortex impairment due to severe cerebral palsy may not be able to use BCIs based on EEG or single-neuron activity from these cortical areas. In such people, BCI systems that use other EEG components or neuronal activity from other brain regions might be good alternatives [18].

3.4.1   Possible BCI Uses

It is important to distinguish between a BCI and its applications. The term *BCI* refers to the system of deep learning that records, analyses, and translates the input (i.e., the user's brain signals) into device commands. In contrast, the term *application* refers to the specific purposes or devices to which the output commands are applied. Recent focus on the real-world applications of BCI technology is speeding the transition of BCI research from the laboratory to clinical products useful in everyday life. Although BCI applications could conceivably be clinical or non-clinical (e.g., computer games), this review discusses clinical applications only [19].

The potential clinical uses of BCIs can be classified as: (1) direct control of assistive technologies; and (2) neurorehabilitation. Since the BCI serves as a replacement of normal neuromuscular pathways, the most obvious BCI applications are those that activate and control assistive technologies that are already in place to enable communication and control of the environment. These applications of BCIs to assistive technology encompass the areas of communication, movement control, environmental control, and locomotion. The possible uses of BCIs in neurorehabilitation have just begun to be explored.

3.4.2   Communication

Communication for people who are "locked in" probably represents the most pressing area in need of intervention with BCI technology. Although other applications are under development, restoring communication has been the main focus of the BCI research community.

Distinguished from one another by the specific electrophysiological features measured, three types of EEG-based BCI systems have been tested in human subjects for the purpose of communication, specifically those based on: 1) slow cortical potentials (SCPs); 2) P300 event-related potentials; and 3) sensorimotor rhythms (SMRs). Both the SCP BCI and the SMR BCI require significant training of the users to gain sufficient control of their brain activity to produce signals that can be effectively applied to BCI use. In contrast, a P300 BCI measures the brain's response to stimuli (visual or auditory) of special significance and requires minimal user training [20].

3.4.3   Movement Control

Restoration of motor control in paralyzed patients is another key application of BCI and is the main goal of many researchers in the field. The research in this clinical application is sparse and has used mainly SMR-based systems. Wolpaw and his colleagues have demonstrated one-dimensional, two-dimensional, and even three-dimensional cursor controls using an SMR system and have done preliminary experiments with SMR control of a robotic arm [21]. These experiments indicate that SMR BCI systems might be able to support multidimensional control of the movement of motor neuroprosthesis or an orthotic device such as a robotic arm. Pfurtscheller and colleagues tested an SMR-based BCI for restoration of motor control in paralyzed patients (for review see *A tetraplegic patient was trained to control an electrically driven hand orthosis with EEG signals recorded over sensorimotor cortex* [22]). By learning to generate separable motor imagery tasks, this patient was able to open and close his paralyzed hand with the hand orthosis.

Functional Electrical Stimulation (FES) can also be used for restoration of motor function in paralyzed patients with intact lower motor neuron and peripheral nerve function. With the goal of further enhancing motor restoration in paralyzed patients, Pfurtscheller and his colleagues combined the SMR BCI with FES systems and tested the combined system in two patients with high spinal cord injury.

3.4.4   Environment Control

BCI-based environmental control could greatly improve the quality of life of severely disabled people. People with severe motor disabilities are often home-bound. Effective means for controlling their environments (e.g., controlling room temperature, light, power beds, TV, etc.) would increase their well-being and sense of independence. A recent pilot study by Cincotti *et al.* attempted to integrate BCI technology into a domestic environmental control system. With unified control through EEG-based BCI technology, the user is able to operate remotely domestic devices such as neon lights and bulbs, TV and stereo sets, a motorized bed, an acoustic alarm, a front door opener, and a telephone, as well as to monitor the surrounding environment through wireless cameras [23]. The clinical validation of the system prototype took place in a simulated home environment in an occupational therapy department. Fourteen healthy normal subjects and four subjects suffering from spinal muscular atrophy type II (SMA II) or Duchenne Muscular Dystrophy (DMD) were tested. The patients were able to control the system with an average accuracy of 60-75% over the last three testing session (8-12 sessions in total). Preliminary findings from this study suggested that the self-control of the domestic environment realized with BCI technology increased the patient's sense of independence. Also, caregivers could be relieved to some extent from the need to be continually present [24].

3.4.5   Locomotion

Restoration of independent locomotion is another important issue for paralyzed people. In light of this, several BCI research groups have attempted to develop BCI-driven wheelchairs in order to restore some form of mobility. Tanaka *et al.* developed an electric wheelchair controlled by EEG. Directional commands were detected by EEG and were then applied to direct control of the wheelchair. Such precise control may be quite demanding on the user [25]. Rebsamen *et al.* reported a wheelchair controlled by a P300-BCI system in which the user simply selects a destination from a menu of destinations. While this approach is less demanding for the user, the capacity for real-time directional control of the wheelchair is limited by the selections and prior definition of possible paths is needed. Millán and his group studied a BCI-controlled wheelchair that is based on the EEG activity associated with various mental tasks and a shared control system. It employed intelligent algorithms to assist the user in obtaining continuous command of the system during wheelchair navigation. Further work is required to confirm the usability of a BCI-driven wheelchair in the real-world environment. For this application, due to considerations of safety, there must be stricter requirements for accuracy than for many other BCI applications.

3.5   Cloud-Based EEG Overview

Cloud computing is the practice of using a network of remote servers hosted on the Internet to store, manage, and process data, rather than on a local server or personal computer. This simply means that data and software are stored on offsite, remote servers, known as a server-farm infrastructure, and accessed over the Internet. In traditional computing, EEG recording and interpretation technology would be found on a local computer's hard drive with networked servers used for archiving and data back-up. With Cloud-based technology, the local data center, as well as offsite data back-up systems, can be eliminated [26].

With Cloud-based EEG technology, the software to perform a study is provided as a service by another company and is known as Software as a Service (SaaS). The patient is set up and the vEEG is recorded in the same manner as it would be on a digital EEG system, but instead of the data being recorded, accessed, and stored on a local system, the software to run the study and the data storage server are seamlessly accessed over the Internet. The server is located in a secure and often unknown location – somewhere on "the Cloud". Once on the Cloud server, all data remains encrypted and is securely protected. With SaaS, software licenses, updates, and security are provided, maintained, and completely managed by IT professionals who specialize in this type of service, relieving the EEG technologist and hospital IT department of this responsibility. The latest versions of software are rolled out to users who then have the option to push the update. Services are available on-demand and are often provided on a "pay-as-you-go" subscription basis. This means that data is always available, wherever the user may be, as long as there is a good Internet connection. Most Cloud providers are extremely reliable in providing services, with many maintaining a 99.9% uptime.

In order to access recorded EEG and video data for pruning or review, a Client Application is installed onto a local computer and "requests" the pages of data from the server. The SaaS opens the files and the Client only allows the data to be viewed.

Data is never downloaded onto a user's computer, therefore, the opportunity for electronic Protected Health Information (ePHI) and patient data to be left on any computer, whether it is open, closed, on or off, is completely eliminated.

Cloud-based computing systems are much more flexible than traditional methods for data recording and storage. Server and infrastructure needs are monitored and completely managed behind the scenes and can expand seamlessly when more space is required. Software solutions are kept current and will not become obsolete with changes in technology.

Cloud solutions can be provisioned for peak times, then deprovisioned when increased capacity is no longer needed. This allows a flexible structure that adjusts according to business demands; costs for use fluctuate and grow with changes in business.

3.5.1   Data Backup and Restoration

HIPAA requires that data is protected using state of the art methods. Physical security of the server location is important, as well as ensuring that environmental conditions are optimal. This includes completely automatic back-up processes, continuous data protection, and frequent back-ups to geographically diverse locations. There should be 24-hour data monitoring, redundant power supplies with strict environmental controls, and verification that the data is being transferred correctly between the client and server. Patient data should never be stored without proper backup. Redundant disk drives, which create full, mirrored backups of the data, are part of the critical process for recovery from corruption or data loss. Geographic redundancy ensures the data is stored in more than one location, so in the event of a catastrophic system failure due to natural or other disasters, the data is still safe, secure, and accessible. Access should be available to data whenever and wherever it is needed, and an export function should be available to download EEG data from the system onto whatever digital medium is preferred by the user, with all access tracked in the audit log [26]. It is imperative that each hospital or user group has a Disaster Recovery Plan in place should data restoration ever be necessary. This documented set of procedures and processes are critical for protecting and restoring an IT infrastructure in the event of a catastrophe.

3.6   Conclusion

An emotion can be identified by extracting different kinds of features from a signal. After preprocessing the signal, it has to be smoothed and optimized for a particular feature, using different deep learning optimization techniques like GA, PSO, etc. After getting the optimized result, it is applied to the Neural Network or SVM and it will provide a high accuracy of emotions in any situation. These emotions will be used for Human Computer Interaction, Affective Computing, Robotics, etc.

References

1. A.Kumar, PS. Rathore, V. Dutt "An IOT Methodology for Reducing Classification error in face Recognition with the Commuted Concept of Conventional Algorithm" has been published in IJITEE and into the press Volume-8 Issue-11, September 2019 ISSN: ISSN:2278-3075
2. Abhishek Kumar & Jyotir Moy Chatterjee & Pramod Singh Rathore, 2020. "Smartphone Confrontational Applications and Security Issues," International Journal of Risk and Contingency Management (IJRCM), IGI Global, vol. 9(2), pages 1-18, April.
3. Asthana, S. Zafeiriou, S. Cheng, and M. Pantic, "Incremental face alignment in the wild," in In Proc. of 2013 IEEE Conference 477 on Computer Vision and Pattern Recognition (CVPR), 2014.
4. Bahareh, Evolutionary Computation Algorithms for feature selection of EEG based emotion Recognition. Elsevier 2017, pp. 143–155.

5. Bhargava, N., Bhargava, R., Rathore, P. S., & Kumar, A. (2020). Texture Recognition Using Gabor Filter for Extracting Feature Vectors With the Regression Mining Algorithm. International Journal of Risk and Contingency Management (IJRCM), 9(3), 31-44. doi:10.4018/IJRCM.2020070103

6. Bozhkov, L., Koprinkova-Hristova, P., Georgieva, P., 2016. Learning to decode human emotions with Echo State Networks. Neural Networks, Vol. 78, p. 112119, June 2016.

7. Bozhkov, L., Koprinkova-Hristova, P., Georgieva, P., Reservoir computing for emotion valence discrimination from EEG signals. Neurocomputing, Vol. 231, p. 2840, March, 2017.

8. C.petrantonais, "Emotion Recognition from EEG using higher order" Vol. 14 pp. 390-396 in 2010 IEEE

9. D. V. Poltavski, "The use of single-electrode wireless EEG in biobehavioral investigations," Methods in Molecular Biology, vol. 1256, pp. 375-390, 2015.

10. F. Silveira, B. Eriksson, A. Sheth, and A. Sheppard, "Predicting audience responses to movie content from electro-dermal activity signals," in Proceedings of the 2013 ACM International Joint Conference on Pervasive and Ubiquitous Computing, 2013, UbiComp '13, pp. 707–716.

11. J. Atkinson and D. Campos, "Improving BCI-based emotion recognition by combining EEG feature selection and kernel classifiers," Expert Systems with Applications , vol. 47, pp. 35–41, 2016.

12. J. Hu, C.S. Wang, M. Wu, Y.X. Du, Y. He and J.H. She, Removal of EOG and EMG artifacts from EEG using combination of functional link neural network and adaptive neural fuzzy inference system, in Neurocomputing, 151.1(0):278–287, 2015

13. J. Katona, I. Farkas, T. Ujbanyi, P. Dukan and A. Kovari, "Evaluation of the NeuroSky MindFlex EEG headset brain waves data," in Applied Machine Intelligence and Informatics (SAMI), 2014 IEEE 12th International Symposium on, Herl'any, Slovakia, 2014.

14. J. Preethi, M. Sreeshakthy, A.Dhilipan, "A Survey on EEG Based Emotion Analysis using various Feature Extraction Techniques", International Journal of Science, Engineering and Technology Research (IJSETR), Volume 3, Issue 11, November 2014.

15. Kumar, A., Chatterjee, J. M., & Díaz, V. G. (2020). A novel hybrid approach of svm combined with nlp and probabilistic neural network for email phishing. International Journal of Electrical and Computer Engineering, 10(1), 486.

16. M. AlzeerAlhouseini, I. F. Al-Shaikhli, A. W. bin Abdul Rahman, and M. A. Dzulkifli, "Emotion Detection Using Physiological Signals EEG & ECG," Journal of Clinical Neurophysiology ,vol. 33, no. 4, pp. 308–311, 2016

17. M. Soleymani, M. Larson, T. Pun, and A. Hanjalic, "Corpus development for affective video indexing," IEEE Transactions on Multimedia, vol. 16, no. 4, pp. 1075–1089, 2014.

18. M.K. Kim, M. Kim, E. Oh, and S.P. Kim, "A review on the computational methods for emotional state estimation from the human EEG," Computational and Mathematical Methods in Medicine, Article ID 573734, 2013.

19. Mehdi Hajinoroozi, Zijing Mao, Yuan-Pin Lin, and Yufei Huang, "Deep transfer learning for cross-subject and cross-experiment prediction of image rapid serial visual presentation events from eeg data," in International Conference on Augmented Cognition. Springer, 2017, pp. 45–55.

20. N. Bhargava, S. Dayma, A. Kumar and P. Singh, "An approach for classification using simple CART algorithm in WEKA," 2017 11th International Conference on Intelligent Systems and Control (ISCO), Coimbatore, 2017, pp. 212-216, doi: 10.1109/ISCO.2017.7855983.

21. Naveen Kumar, Prakarti Triwedi, Pramod Singh Rathore, "An Adaptive Approach for image adaptive watermarking using Elliptical curve cryptography (ECC)", First International Conference on Information Technology and Knowledge Management pp. 89–92, ISSN 2300-5963 ACSIS, Vol. 14 DOI: 10.15439/2018KM19

22. R. M. Mehmood and H. J. Lee, "Exploration of Prominent Frequency Wave in EEG Signals from Brain Sensors Network" International Journal of Distributed Sensor Networks, 2015.

23. Rathore, P.S., Chatterjee, J.M., Kumar, A. *et al.* Energy-efficient cluster head selection through relay approach for WSN. J Supercomput (2021). https://doi.org/10.1007/s11227-020-03593-4

24. S.-K. Tai, C.-Y. Liao and R.-C. Chen, "Exploration of Multi-Brainwave System Mainframe Design," ICIC Express Letters, Part B: Applications, vol. 8, no. 9, pp. 1307-1314, 2017.

25. Singh Rathore, P., Kumar, A., & Gracia-Diaz, V. (2020). A Holistic Methodology for Improved RFID Network Lifetime by Advanced Cluster Head Selection using Dragonfly Algorithm. International Journal Of Interactive Multimedia And Artificial Intelligence, 6 (Regular Issue), 8. http://doi.org/10.9781/ijimai.2020.05.003

26. Y. Lv, Y. Duan, W. Kang, Z. Li, F. Y. Wang, "Traffic Flow Prediction With Big Data: A Deep Learning Approach" IEEE Transaction on Intelligent Transportation Systems, vol. 16, no. 2, pp. 865-873, 2015.

4

Effective and Efficient Wind Power Generation Using Bifarious Solar PV System

R. Amirtha Katesa Sai Raj*, M. Arun Kumar, S. Dinesh, U. Harisudhan and Dr. R. Uthirasamy

Department of EEE, KPR Institute of Engineering and Technology, Coimbatore, India

Abstract

Electricity is one of the most essential requirements in day-to-day life. Generally, electricity is generated through renewable and non-renewable energy sources. Power generation through the burning of coal is effective, but it also causes pollution and reduces the quantity of fossil fuels, which leads to degradation of natural resources. Nowadays, renewable energy sources are widely used because they are green and used to generate power without degrading any natural resources. Solar and wind energy are two of the most effective and efficient renewable energy sources, but windmills can produce power only when sufficient wind is blowing and a solar PV system can generate power only when there is sunlight. These two limitations degrade the efficiency of prolonged power generation. The main motivation of the proposed work is to improve the efficiency of wind power generation with the use of solar panels and utilize the power generated by solar cells effectively by powering the electrical components used inside a windmill, such as the revolving motor, elevators, etc. In this chapter, various literatures have been reviewed and the remarkable features of the proposed system are highlighted. At the end, data analysis is done using Deep Learning and visualizes all the results in graphical form. In this analysis, the power generation from our proposed system and traditional methods is visualized.

Keywords: Solar panel, DC motor, cut-in speed, battery, yaw motor

Corresponding author: sairaj2306@gmail.com

Pramod Singh Rathore, Vishal Dutt, Rashmi Agrawal, Satya Murthy Sasubilli, and Srinivasa Rao Swarna (eds.) Deep Learning Approaches to Cloud Security, (43–62) © 2022 Scrivener Publishing LLC

4.1   Introduction

In our world, we are facing many shortage crises, including water, natural resources, and power demand. These crises will lead the human habitat to a dangerous situation. This is due to the ever-growing population. One of the main requirements for our day-to-day life is electricity.

Power demand in our country increases rapidly and the power demand of Tamil Nadu is 50 Giga-watts. In order to meet this demand, renewable energy must be utilized efficiently and effectively [1].

The main objective of the proposed work is to make use of wind power effectively and efficiently by using solar panels. Various concepts and ideas have been presented for placing solar panels in windmills, but they are not efficient. The proposed system implies an effective and efficient design for placing solar panels in windmills. The concept also implies running the rotor blades under a lesser cut-in speed by using a DC motor coupled to the gearing arrangement which is powered by solar power.

Looking at the Figure 4.1, clearly shows how important the generation of wind power is across many locations, so the main idea is to use solar power in the wind energy conversion system to ensure there is continuous generation of wind power even when there is an absence of

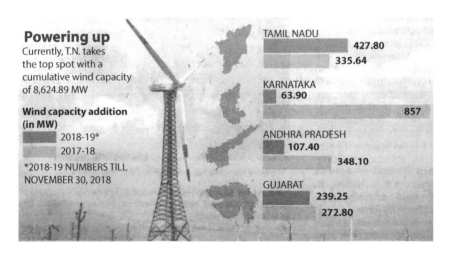

Figure 4.1 Survey of WECS.

wind. This solar power is generated using a **Bi-Facial Solar Panel** and it gives power in a much more efficient manner than the conventional panel [2].

4.2   Study of Bi-Facial Solar Panel

Bifarious (i.e., Bi-facial) solar panels Figure 4.2, are one of the most effective and efficient sources of solar power generation. They have solar PV cells in the front side of the panel, as well as on the back. This construction of PV placement generates effective power generation is small sizes.

An important feature of these panels is that they can be placed in small areas yet can deliver two times the power of conventional panels.

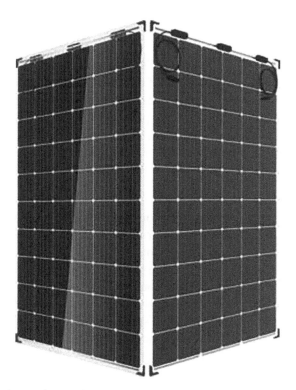

Figure 4.2 Bi-facial solar panel.

Conversely, tthe cost of bi-facial panels is higher when compared to conventional panels.

Panels are placed on top of the nacelle in a spoiler-like structure and also on the tower in a circular manner at equal intervals [3].

4.3   Proposed System

The proposed system is used to improve the power generation of the wind energy conversion system by the use of bifarious solar panels. Power generated from the solar panels is given to the battery, yaw motor, and DC motor when supply to them is needed, all other power from the solar panels is given to the grid via inverter.

Li-ion batteries are used to store power during the daytime and stored power is given to the electrical components at night. The DC motor coupled to the gearing system is powered by solar energy and is used to give a starting push to the rotor blades when the wind speed is a little less than the cut-in speed [4].

4.3.1   Block Diagram

The block diagram representation of the proposed system is shown in the Figure 4.3. Daytime and nighttime arrow marks represent power generation and utilization at different time periods [5].

The system consists of solar panels, a charge controller, battery bank, yaw motor, DC motor, gearbox, rotor blades, and an inverter unit with its controller unit. Typical working of the proposed system is explained as follows:

- Solar panels are placed on top of the nacelle as well as on the tower to generate power during the daytime and supply it to various electrical components in the WECS.
- Excess power is given to the grid via an inverter and charges the battery bank simultaneously.
- The DC motor is operated when the cut-in speed of the wind is a little less than its rated speed, so even at low speed the WECS will be able to generate rated power.
- At night, the battery bank acts as a power source to the electrical components [6].

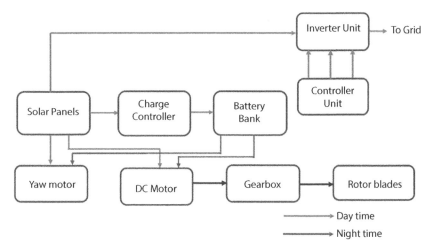

Figure 4.3 Block diagram of the proposed system.

4.3.2 DC Motor Mechanism

A DC motor is used as a revolving element, which is used to rotate the rotor of the wind turbine when the cut-in speed of the wind is a little less than the rated speed. Since the ratio of the turbine is greater, a high starting torque is required to give a push to the rotor blades in order to gain momentum, so a DC series motor is used [7].

The motor is placed in the gearing system, which drives the rotor of the wind turbine. Gearing arrangement is housed inside the nacelle and the motor is coupled to it as shown in the Figure 4.4.

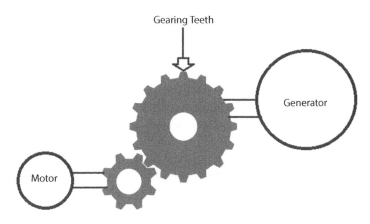

Figure 4.4 DC motor with gear arrangement.

When there is no sufficient wind to generate power, the turbine does not rotate; thee cut-in speed of the wind is less and solar power is given to the DC motor to rotate the rotor. The DC motor gives a startup push to the turbine and it starts rotating, generating power at lesser speeds [8].

The main advantage of using a DC motor is that there is no need for an inverter circuit, so there is no need to use an AC motor. Also, a DC series motor has a high starting torque when compared to AC motors.

An AC induction motor can be also used, but by using an induction motor, the reactive power must be compensated for or efficiency will be reduced. Another drawback of using an AC motor is its maintenance.

The DC motor is ideal when there is sufficient wind flowing and the power generated from solar energy is given to the grid via an inverter. Use of panels on towers also increases the power generated from solar energy.

The power supply to the DC motor is from the solar panels during daytime and from battery banks at night [9].

4.3.3   Battery Bank

Li-ion batteries are used as backup power when there is no sunlight. Li-ion batteries are very efficient because of their less ideal discharge rate. Multiple batteries are taken and connected to form a stack of batteries. Battery banks are then placed inside the nacelle [10].

4.3.4   System Management Using IoT

The term Internet of Things (IoT) normally refers to circumstances where network connectivity and computing potential extends to objects, sensors, and everyday items not normally measured by computers, allowing these devices to generate, exchange, and consume data with minimal human intervention. There is, however, no single common definition. The recent convergence of several technology market tendencies, however, is bringing the Internet of Things closer to widespread reality. This includes ubiquitous connectivity, widespread adoption of IP-based networking, computing economics, miniaturization, advances in data analytics, and the rise of cloud computing.

An IoT based network has been established to enhance the following functionalities in the proposed WECS:

1. To monitor the battery management system

2. To initiate the automation process between the WECS, battery bank, solar panels, and grid system
3. To monitor the fault occurrence in the wind turbine, solar panels, and battery management system

A fault is defined as the extinction of the ability of a system to complete a function. When a fault, such as abnormal status in speed, temperature, voltage, current, or vibration, occurs inside the wind turbine, a separate system is to be taken care for further processing. It catalogs the consequence of the fault and responds referring to the type of the break down. In order to keep away from safety hazards or main system breakdowns, the turbine has to be shut down. Often, they are resumed because of wrong failure detection which could be caused by noise within the system and, therefore, these faults are not considered crucial problems. If the failure is severe, an illustration examination has to be made which can be carried out by the operators or authorized personnel. Finally, whenever a major breakdown has occurred a report is documented. High speed and low speed shaft faults are the most common failures in a wind turbine. Mostly, the faults in a wind turbine can be detected by current measurements [11].

Similarly, an IoT system will monitor the performance of solar panels placed on the nacelle and tower of the WECS. Monitoring systems supply prosperity of data about the performance and operation of a solar system. Sensors help to monitor external conditions such as:

- Wind speed
- Radiance of the sun
- Temperature
- Cell temperature

They also help measure the parameters of the surrounding elements:

- String voltage
- Inverter performance
- Energy production
- System output (AC electric energy)
- Battery voltage levels to determine battery health

The charge and discharge rates of batteries are monitored and the information can be tracked at any time by the user by use of IoT. The charging rate of the battery is sent to the user to track whether the battery is charging or not.

The discharge rate of the battery is also monitored so that the user can tell the amount of charge left in the battery. The discharge rate is monitored by the power drawn from the various electrical components in the WECS (i.e., DC motor, yaw motor).

Similarly, the charging rate of the batteries can be monitored during the daytime when the batteries are powered by solar energy. Getting to know these details, the user can effectively utilize the amount of charge available in the battery. This information can be sent to a user by means of SMS.

The IoT is executed by interfacing wireless sensors to the Internet by appropriate hardware interface which will pass enormous data to the operative in a far place. Consequently, by making an allowance for all the presentation parameters and conditions, an appropriate control mechanism can be agreed upon via Internet to the microprocessor to manage all the devices according to our desired manner [12].

The IoT plays a major role in the grid connected WECS. It makes the WECS smarter with the enabled sensors. In the proposed system excess power generated by solar panels is exported to the grid. This process has been monitored and regulated by IoT based automation systems.

4.3.5   Structure of Proposed System

1. Wind Blades
2. Nacelle
3. Gear
4. Generator
5. Spoiler Design
6. Bi-facial solar panels
7. Battery Bank
8. Yaw system
9. Tower
10. Support
11. Solar panels
12. Revolving Motor

Figure 4.5 represents the diagrammatical structure of the proposed system. It consists of the components that are used in the proposed system.

The working of the proposed system is based on combining solar and wind together. Solar panels are placed on top of the nacelle in a spoiler like

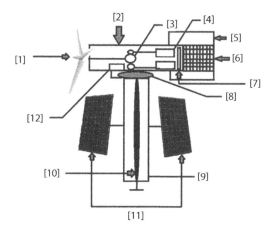

Figure 4.5 Structure of proposed system.

design and are also placed on the towers of the wind turbine. The generated solar power is given to the grid via inverter and is also used to power the DC motor and the revolving motor in the wind turbine [13].

4.3.6 Spoiler Design

A spoiler as shown in Figure 4.6, structure is used to reduce the airflow falling on the panels that are placed on top of the nacelle. Its structure is composed in the nacelle to reduce the weight falling on the nacelle. This structure is mainly used to support the aerodynamics.

Figure 4.7 shows the placing of a solar panel inside the spoiler structure that is placed on top of the nacelle. When there is sufficient wind flowing to rotate the rotor, solar power is stored and directly given to the grid via inverter. When there is no wind flowing, the solar power is given to the DC motor [14]. The DC motor is placed in the gearing system and rotates the rotor of the wind turbine. Li-ion batteries power the DC motor at night and when solar energy is not dominant. The revolving motor of the yaw system is also powered by solar power and rotates according to the direction of the wind. By this technique, the power generation of wind energy will be increased and the overall efficiency of the system will increase [15].

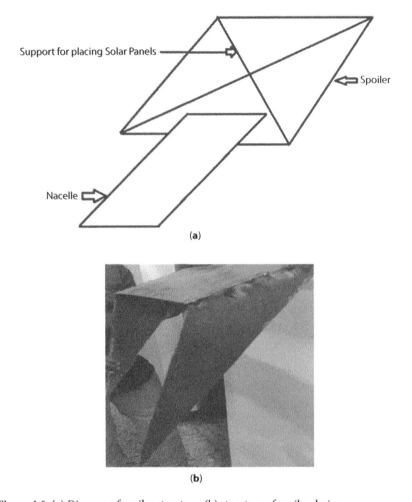

Figure 4.6 (a) Diagram of spoiler structure, (b) structure of spoiler design.

4.3.7   Working Principle of Proposed System

The working of the proposed system is based on a combination of solar and wind energy together. Bi-facial solar panels are placed on top of the nacelle in a spoiler-like design and are also placed on the tower of the wind turbine.

The generated solar power is given to the grid via an inverter and is also used to power the DC motor and the revolving motor in the wind turbine [16].

When there is sufficient wind flowing to rotate the rotor, the solar power is stored and directly given to the grid via an inverter and when the wind flowing is less than normal, solar power is given to the DC motor.

Figure 4.7 Snapshot of bifarious solar panel in spoiler structure.

The revolving motor of the yaw system is also powered the solar power and rotates according to the direction of the wind. By this technique, the power generation of wind energy will be increased and the overall system efficiency will be increased.

4.3.8 Design and Analysis

Parameters	Rating
Solar Panels	60 Watts
DC Motor	12V, 9600rpm
Yaw Motor	12V, 9600rpm
Battery	22V, 15600mAh
Generator	20 Watts

4.4 Applications of IoT in Renewable Energy Resources

The Internet is a gigantic intercontinental network of networks. The modern node is the Internet of Things (IoT). This technique is widely used in many industries as it can be accessed from any place. The user can enjoy the services of these things remotely. The main advantage is that it can be controlled remotely, via the internet. Similarly, the Internet of Things (IoT) plays a vital role in the management of renewable energy resources. The experimental setup of the proposed system is shown in Figure 4.8.

Figure 4.8 Experimental setup of the proposed system.

4.4.1 Wind Turbine Reliability Using IoT

To provide a real-time monitoring of wind turbine farms and collect data wirelessly, each turbine within the array can be equipped with its own IoT device. These IoT platforms will be helpful in determining the power output model of the wind turbine, called a power curve. It will also be helpful in predicting the lifetime of the battery used in the wind turbines. With the help of IoT, various applications can be employed, such as constituting the yaw angle and pitch angle in a periodic fashion.

The reliability of wind turbines supports a sustainable operation of such an important power source. This mandates a real-time monitoring system of wind turbine conditions to reduce human intervention as much as possible. The IoT provides a promising solution to providing an online monitoring system and control of the operation of the sensors and actuators of the wind turbine.

4.4.2   Siting of Wind Resource Using IoT

IoT plays a significant role in selecting the most suitable location for installing wind turbines. Meteorological sensors will respond to real-time weather conditions and are designed to monitor specific environmental parameters, such as wind speed, wind direction, and ambient direction.

The main factor of importance to the wind turbine is the speed of the wind, which is responsible for the power generation. Thus, wind speed and monthly average wind speed can be predicted as shown in Figure 4.10 and 4.11 respectively.

Wind turbines can be monitored through sensors to track air flow and mechanical power that will be extracted from wind. For sensors, factors such as highness, location, accuracy, duration, quality, and quantity should be considered as shown in Figure 4.9 in the form of features of IoT. Moreover, better IoT performance depends on trusted and identified measurements including data sampling, recording, handling, recovery, quality, quantity, and format, which should all be taken into account. We have carried out these steps using the IoT to maintain reliability performance within a scheduled plan. As sensor technology advances, it becomes

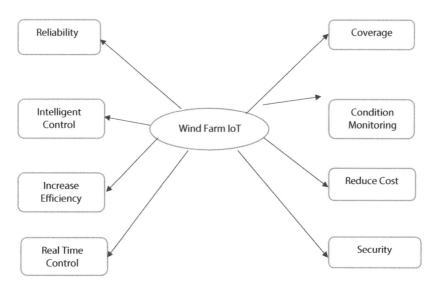

Figure 4.9 IoT features.

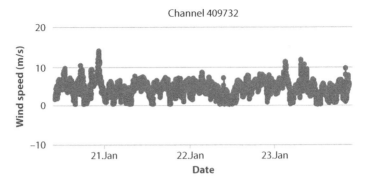

Figure 4.10 Wind speed using IoT.

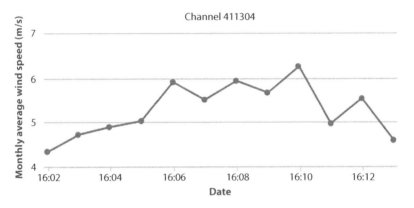

Figure 4.11 Monthly average wind speed using IoT.

possible to monitor wind turbines on wind farms, ensuring maximization of air flow and mechanical power [17].

The turbulence intensity (TI) indicator is a very important factor for wind turbine siting. The TI normalizes the standard deviation of wind speed with the mean wind speed. Thus, intensity plays a major role in calculating the wind speed.

4.4.3   Application of Renewable Energy in Medical Industries

Studies have shown that the share of hospitals in energy consumption of the building sector is ~6%. Hospitals and clinics have several facilities to provide required energy for heating, cooling, and hot water preparation, etc. It shows energy balance in a hospital. Reduction in energy consumption of hospitals gained importance in the goal to achieve green hospitals. Various solutions, including regular maintenance of filters utilized in heating, ventilation, and

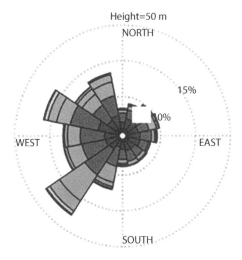

Figure 4.12 Wind rose.

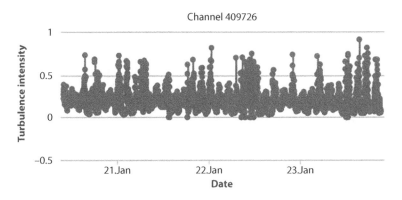

Figure 4.13 Turbulence intensity using IoT.

air conditioning (HVAC) systems, using isolating materials, applying aquifer systems, and improving air conditioning systems, are represented to reduce energy consumption in hospitals in the form of wind rose as shown in Figure 4.12 and turbulence intensity as shown in Figure 4.13 [18].

4.4.4   Data Analysis Using Deep Learning

The power generated from the above proposed system has to be compared to the traditional system. This process is necessary to understanding the importance of the proposed model, its accuracy, and its pros and cons. To find out these parameters, we use techniques of Deep Learning that

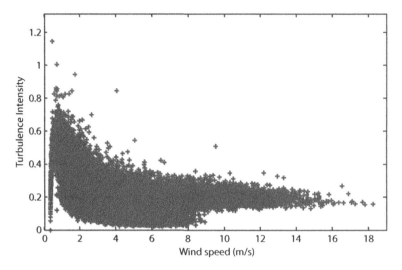

Figure 4.14 Turbulence intensity vs. wind speed (m/s).

provide easier and more convenient ways of gathering data. Deep Learning is widely used in various domains like the medical and electrical domains to analyze data. Electrical equipment can generate electricity or power, but load management in power grids can be managed using computer-oriented methods and techniques. Deep Learning Neural Networks play a vital role in the area. Due to the technology, the concept of renewable energy is made possible [19].

To analyze the data, we have various computational methods in a Deep Learning Neural Network which can learn from the data generated and make predictions for the future. We will compare two techniques of Deep Learning to get a better performance algorithm to predict the outcomes.

The Gradient Boosting and XG Boost techniques will be incorporated to fulfill the task as shown in Figure 4.15.

Step 1: Data Preprocessing: This step includes data gathering from the proposed model. The data will be collected in segments of 1 hour. The data may include wind power, turbulence intensity as shown in Figure 4.14, etc.

Step 2: Cross-Validation: After gathering the data, the Deep Learning algorithms create a model. After building a model, we have to select the best hyper-parameter values from the model. In this direction, we will divide the first segment data into three parts: Training Set, Validation Set, and Test Set.

Step 3: Apply Step 1 and Step 2 for each segment.

Step 4: Subset Selection: After Step 3, we have to select the most appropriate subset from the models.

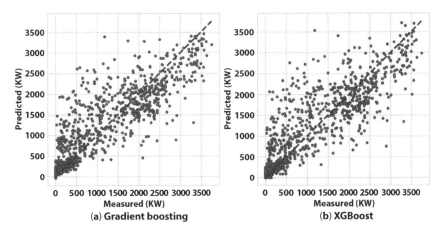

Figure 4.15   Comparison of gradient boosting and XG boost.

4.5   Conclusion

A combination of wind and solar power will produce energy effectively and improve the efficiency of a system. This concept produces prolonged power generation and the use of renewable energy will be utilized effectively, ensuring green energy. This system is also useful for a continuously reliable power supply. The main objective of the proposed system is to increase the complete utilization of power generated by WECS and promote the use of green energy. After using comparison techniques of Deep Learning, we can find which one is the best for analyzing the power generated from the proposed model.

References

1. Abhishek Kumar & Jyotir Moy Chatterjee & Pramod Singh Rathore, 2020. "Smartphone Confrontational Applications and Security Issues," International Journal of Risk and Contingency Management (IJRCM), IGI Global, vol. 9(2), pages 1-18, April.
2. Bhargava, N., Bhargava, R., Rathore, P. S., & Kumar, A. (2020). Texture Recognition Using Gabor Filter for Extracting Feature Vectors With the Regression Mining Algorithm. International Journal of Risk and Contingency Management (IJRCM), 9(3), 31-44. doi:10.4018/IJRCM.2020070103.

3. Binayak Bhandari (2016) 'Optimization of Hybrid Renewable Energy Power System', International Journal of Precision Engineering & its Applications Vol.17, No.6.

4. Gwani, Muzamil and Taheri (2014) 'Urban Eco-Green Energy Hybrid Wind-Solar Photovoltaic Energy System & its Applications', International Journal of Precision Engineering & Manufacturing Vol.16, No.5.

5. Hughlett and Mike (2019) 'Minnesota Wind-Solar Hybrid Project could be New Frontier for Renewable Energy', Star Tribune Archived from the original on Vol.2, No.9.

6. Jossi and Frank (2019) 'Wind-Solar Pairing Cuts Equipment Costs while Ramping up Output Renewable Energy Network', Energy News Network. Archived from the original on Vol.18, No.19.

7. Kumar, A., Chatterjee, J. M., & Díaz, V. G. (2020). A novel hybrid approach of svm combined with nlp and probabilistic neural network for email phishing. International Journal of Electrical and Computer Engineering, 10(1), 486.

8. N. Bhargava, A. Kumar Sharma, A. Kumar and P. S. Rathoe, "An adaptive method for edge preserving denoising," *2017 2nd International Conference on Communication and Electronics Systems (ICCES)*, Coimbatore, 2017, pp. 600-604, doi: 10.1109/CESYS.2017.8321149.

9. N. Bhargava, S. Dayma, A. Kumar and P. Singh, "An approach for classification using simple CART algorithm in WEKA," *2017 11th International Conference on Intelligent Systems and Control (ISCO)*, Coimbatore, 2017, pp. 212-216, doi: 10.1109/ISCO.2017.7855983.

10. Naveen Kumar, Prakarti Triwedi, Pramod Singh Rathore, "An Adaptive Approach for image adaptive watermarking using Elliptical curve cryptography (ECC)", First International Conference on Information Technology and Knowledge Management pp. 89–92, ISSN 2300-5963 ACSIS, Vol. 14 DOI: 10.15439/2018KM19.

11. Palash Jain (2016) 'Performance Prediction and Fundamental understanding of Small Scale Vertical Axis Wind Turbine with Variable Amplitude Blade Pitching', Renewable Energy Vol.9, No.7.

12. Perlin John (2017) 'From Earth: The Story of Solar Electricity Space', to International Journal of Research in Advent Technology, Vol.5, No.5.

13. Rathore, P.S., Chatterjee, J.M., Kumar, A. *et al.* Energy-efficient cluster head selection through relay approach for WSN. J Supercomput (2021). https://doi.org/10.1007/s11227-020-03593-4.

14. Siavash Taghipour Broujeni (2013) 'Hybrid PV/Wind Power System Control for Maximum Power Extraction and Output Voltage Regulation', Amirkabir University of Technology Tehran, Iran Vol.4, No.5.

15. Singh Rathore, P., Kumar, A., & Gracia-Diaz, V. (2020). A Holistic Methodology for Improved RFID Network Lifetime by Advanced Cluster Head Selection using Dragonfly Algorithm. International Journal Of Interactive Multimedia And Artificial Intelligence, 6 (Regular Issue), 8. http://doi.org/10.9781/ijimai.2020.05.003.

16. Sunanda Sinha (2015) 'Prospects of Solar Photovoltaic Micro-Wind Based Hybrid Power System', Western Himalayan State of Himachal Pradesh in India, Energy Conversion and Management Vol.3, No.3.
17. Vikas Khare, Nema, Savita and Baredar (2017) 'Bifacial Solar-Wind Hybrid Renewable Energy System' A Review, International Journal of Research in Advent Technology, Vol.2, No.5.
18. Bhagwat, Ajay & Teli, S. & Gunaki, Pradeep & Majali, Vijay. (2015). Review Paper on Energy Efficiency Technologies for Heating , Ventilation and Air Conditioning (HVAC). International Journal of Scientific & Engineering Research 6.
19. Shaikh, Mohd Rizwan & Shaikh, Sirajuddin & Waghmare, Santosh & Labade, Suvarna & Tekale, Anil. (2017). A Review Paper on Electricity Generation from Solar Energy. International Journal for Research in Applied Science and Engineering Technology. 887 f10.22214/ijraset.2017.9272.

5

Background Mosaicing Model for Wide Area Surveillance System

Dr. E. Komagal

Department of Electronics and Communication Engineering,
Latha Mathavan Engineering College, Alagarkovil, Madurai, India

Abstract

Exploiting Pan/Tilt/Zoom (PTZ) cameras permits widening of the field of view of a surveyed area, but it needs a more efficient computational cost method to be employed to detect motion. Most feature-based techniques lead to significant redundancy, having large frame intervals for feature correspondence and a scarcity of corresponding feature points. To overcome the issues of features, the Mosaic Background Model using the SURF feature of Deep Learning is proposed. Mosaicing is a method of assembling multiple overlapping images of the same scene into a larger wide image of a scene which overcomes the above issues. In real-time monitoring, a major problem is that the field of vision is completely too small to capture the target and a larger field of vision with low resolution. Ability to handle all the above issues, which includes management of quantity and quality feature extraction, is proposed. In this paper, Speeded Up Robust Features (SURF) are used to construct a Mosaiced Background Model for foreground segmentation through Deep Learning. The proposed algorithm is able to detect and represent foreground objects and can also efficiently deal with large zoom effect, compared to Scale-Invariant Feature Transform (SIFT) based mosaiced foreground segmentation.

Keywords: PTZ, mosaicing, SURF, video surveillance, feature extraction

Email: komagale@gmail.com

Pramod Singh Rathore, Vishal Dutt, Rashmi Agrawal, Satya Murthy Sasubilli, and Srinivasa Rao Swarna (eds.) *Deep Learning Approaches to Cloud Security*, (63–74) © 2022 Scrivener Publishing LLC

5.1   Introduction

Video surveillance is a very active computer vision research area and the main tasks of a video surveillance system include motion detection, object classification, tracking, understanding of activity, and behavior analysis [1]. Panoramic representations of visual scenes have a wide application scope, including virtual reality (VR), interactive 2D/3D video, tele-conferencing, content-based video compression and manipulation, and full-view video surveillance. A wide Field of View (FOV) lens (e.g., a fisheye) can be a solution for obtaining panoramic views. However, in addition to the high cost of these specially designed image sensors, images obtained by such sensors have substantial distortions and mapping an entire scene into the limited sensor target of a standard video camera has low resolution. Constructing a panoramic representation by mosaicing image sequences captured by PTZ (Pan, Tilt, and Zoom) cameras meets the requirements of the aforementioned applications for high image resolution. Recently, sensor technology developments have led to an increasing distribution of a specific type of moving camera like pan-tilt-zoom (PTZ) cameras that can dynamically modify the field of view through the use of panning, tilting, and zooming (i.e., moving left and right, up and down, and closer and farther away). However, very few of the current algorithms and systems are able to properly detect and represent foreground objects and to efficiently deal with a large zoom effect. Still, this problem exists, which is very important to analyzing video further. We have various Deep Learning approaches that can do the analysis and find the appropriate and desired result for us. This motivates our work described in this paper.

The organization of the paper is given below. The next section describes related works. Section 5.3 explains the overall methodology. Section 5.4 presents the results and discussion, conclusion, and future work.

5.2   Related Work

Many researchers have attempted to extend stationary background modeling methods for PTZ camera-based application. The method presented uses the information of the overlapped region between consecutive frames to generate a background. The limitation of this frame-to-frame method is that it can accumulate errors over a long sequence and cannot zoom in to get a high quality view of the target. Some other methods use a panorama or a panoramic pyramid for background subtraction. However, this method cannot provide a robust background for foreground detection because the panoramas contain moving objects which cause false detection [2].

The timely replacement algorithm based on the central region of the background and the control strategy of camera rotation could achieve target detection and tracking under a dynamic background. The target which has to be tracked was in the field of view which was set before. Currently, this method can only achieve the detection and tracking of a single target, which is the main disadvantage of this paper. The multi-layered correspondence propagation method registers observed frames in the panoramic background and obtains the correspondence background, even when the zoom value (scale) of the PTZ camera changes a lot. Using the correspondence background, foreground objects are detected. Background registration methods are based on phase correlation, block matching, background motion compensation, and so on. Although these algorithms are able to achieve the target detection and tracking of dynamic surroundings, the background matching algorithm is more complex due to the tracking of larger amounts of calculation and lower speeds in real-time processing [3].

Existing feature-based methods often filter the sequence images to extract feature points and try to find matches between a set of points. Subsequently, the tomography of the images will be obtained. The mosaic's performance depends on the corresponding correction very deeply [4]. In order to increase the corresponding correction rate, the feature points are tracked frame-to-frame. On the other hand, video frames are typically 30 frames per second and long sequences contain significant redundancy. Therefore, to meet the real time constraint for the whole system, as a first step, some measures to identify key frames, which includes enough effective information for later mosaicking, is needed. In order to overcome the problems existing in traditional methods, a Background Mosaicing Model for wide area surveillance using SURF features is proposed. The objective of this paper is to propose an effective background model for a PTZ camera-based surveillance system which constructs mosaic and is able to properly detect and represent foreground objects even though areas with large zoom. The proposed method works well in workplaces such as Healthcare and monitoring in public places.

5.3   Methodology

The main aim of this paper is to propose an effective background model for a PTZ camera-based surveillance system that performs feature extraction followed by mosaic construction and foreground segmentation. Figure 5.1 shows an overview of the methodology.

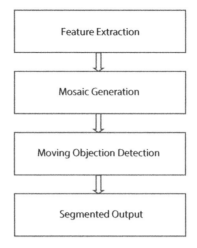

Figure 5.1 Proposed methodology.

Input video is converted into a sequence of frames. Subsequently, the features are extracted using SURF (Speeded Up Robust Features). The first frame is taken as a reference frame and the feature points are compared to the points in the immediate next frame. If the feature points of both frames match, it will be compared to the next successive frame. Tracking these N feature points frame-to-frame is continued up until a mismatch occurs [5]. Then, the current frame will be taken as the key frame. Frames taken from reference frame to key frame will be mosaiced to a single frame (updated reference frame) which contains all the essential information about all the in-between frames optional key frames. Consequently, the key frame is updated as the new reference image and N feature points are selected on it [6]. Finally, the Background Mosaic Construction Model is fed for fore-ground segmentation. The details on methodology are given below.

5.3.1   Feature Extraction

Deep Learning has a vast repository of algorithms and techniques that can help us for feature extraction. PTZ camera-based video is taken for motion segmentation. Video formats, like .avi files, are converted as frames [7]. These frames are provided to the Deep Learning algorithm as an input, then all the converted frames as input data are used for feature extraction. All the converted frames are given for SURF feature Extraction. SURF (Speeded-Up Robust Features) features are based on a detector-descriptor scheme. The detector is based on the Hessian Matrix and uses a very basic approximation just as DoG, which is a Laplacian-based detector. A 'Fast-Hessian' detector relies on integral images to reduce computation time. The descriptor also describes a distribution of Haar-wavelet responses within the interest point neighborhood [8]. Again, exploit integral images for speed. Moreover, only 64 dimensions are used, reducing the time for feature computation and increasing the robustness simultaneously. A new indexing step, based on the sign of the Laplacian, increases the strength of the descriptor. The concept of integral images allows faster implementation of box type convolution filters [9]. The entry of an integral frame ($I\Sigma$ (x)) at a specific location (x = (x, y)) represents the sum of all pixels in the input image (I) of a rectangular region created by the point (x) and the origin shown in Equation (5.1),

$$I_{\Sigma}(X) = \sum_{i=0}^{i \leq x} \sum_{j=0}^{j=y} I(i, j) \tag{5.1}$$

With I_{Σ} calculated, it only takes four additions to determine the sum of the intensities over any upright, rectangular area, independent of its size.

The extracted features are used to construct a Mosaiced Background Model [10].

5.3.2   Background Deep Learning Model Based on Mosaic

Mosaicing consists of four steps:

1. First, detect N feature points on the initial frame of the video and set this frame as the reference frame
2. Then, track these N feature points using SURF algorithm frame-to-frame until condition 1 is broken and set the current frame as a key frame for mosaicing
3. Next, the reference image, the key image, and their corresponding feature points will be sent to the quality feature points management module to obtain the most accurate mosaic according to the condition 2 [11]
4. Lastly, update the key frame as the new reference frame and reselect N feature points on it

The objective of this paper is to propose a Mosaic Based Background Model for quantity management for robust feature points. Here, the speed and performance of the surveillance system will be increased [12].

1. Extract N feature points on the initial frame (F1) of the video.

2. Track these N feature points frame-to-frame until M feature points have been matched and the current frame is F2. M is the threshold for re-selecting new feature points.
3. Once F2 is reached, decide on F1 and F2 as the key frames which have enough information including the sequence F1~F2.
4. Using these two key frames, F1 and F2, and their matched feature points (M), the transform matrix is obtained for mosaicing construction [13].

In this paper, feature points are according to the gradient value. Meanwhile, in order to ensure accuracy of the construction, inputs F1, F2, and their feature points (M) will be subject to the quality management which will be described later. Furthermore, considering that if some strong feature points occur in a video and they are almost matched very well during many frames, step 2 is very difficult to reach and, thus, the interval between F2 and F1 is very large. In order to ensure that the key frames retain enough information, give an upper limitation (H) for the frame interval. If the frame interval between the two key frames is larger than the H, select these two frames as the key frames. Here, a general idea is proposed to identify the upper limitation (H) in different situations [14]. If video frames are typically 30 frames a second, then this long sequence contains significant redundancy. The background model for mosaicing, constructed as shown in Figure 5.2, was implemented on MICC datasets of video frames consisting of 50 frames a second. A mosaiced view of the MICC dataset is shown in Figure 5.2.

The sequence of frames has feature points for each frame generated from SURF and when the sequence gets changed, it should be automatically

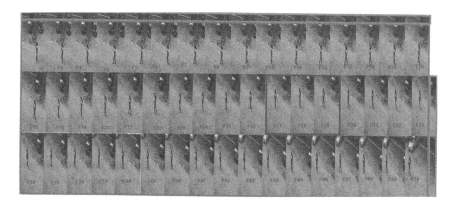

Figure 5.2 MICC dataset background mosaic model.

updated for a mosaic background. Figure 5.2 illustrates how mosaicing is generated for the 50 frames and it indicates a Red as Necessary Frame, Green as Lost Area, and Blue as Overlapping Area [15].

Overlapping area between two key frames should retain enough information for this scene; meanwhile, the lost information must be smaller than a limit. Equations (5.2), (5.3), (5.4), (5.5), and (5.6) are used to define the conditions.

$$\text{Overlap area} > P_{overlap} \tag{5.2}$$

$$\text{Lost area} < P_{lost} \tag{5.3}$$

Overlap is the threshold which is used to make sure that the two key frames' overlap area is large enough [16]. This parameter is often set to 30%. At the same time, the lost information should not be too large to influence the final mosaic significantly. Therefore, another parameter, Plots, is also very useful. This threshold, Plots, is often been set to 10%. Where W is the sum number of Pixels [17],

$$\text{Overlap Area} = \frac{Area(A \cap C)}{Area(oneframe)} \tag{5.4}$$

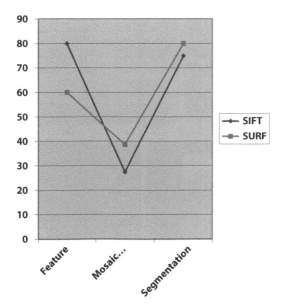

Figure 5.3 Comparison of mosaic features.

$$\text{Lost Area} = \frac{Area\,(A \cup B_1 \cup B_n \cup C) - Area(A \cup C)}{Area\,(oneframe)}$$

(5.5)

$$\text{Difference}\,(X, Y) = \frac{1}{W}\Sigma\|X - Y\|^2$$

(5.6)

5.3.3   Foreground Segmentation

The constructed mosaiced background model as shown in Figure 5.3, is fed to the foreground segmentation. Fixed thresholds for finding the foreground and background are determined. If the threshold is higher than the key frame, then it is foreground and otherwise, background [18].

5.4   Results and Discussion

Foreground segmentation in image sequences taken from PTZ cameras has been experimented with using MATLAB® 2016 software on datasets,

(a)   (b)

(c)   (d)

Figure 5.4 (a) Input frame 220 on VIP dataset, (b) Extracted SURF features, (c) Mosaic background modelling and (d) Foreground segmentation.

such as MICC and VIP Videos, by applying Deep Learning techniques. The sample frames are shown in Figure 5.4 (a-d) and Figure 5.5 (a-d). Input video is used for feature extraction by SURF as shown in Figure 5.4 (a-d) and Figure 5.5 (a-d) and illustrates the robust features [19]. The first frame is taken as a reference frame and is compared to each successive frame. Track these N feature points using a SURF algorithm frame-to-frame until feature points mismatch. Then, set the current frame as a key frame for mosaicking, updating the key frame to the new reference image and selecting N feature points on it. Finally, the Background Mosaic Construction is made as shown in Figures 5.4 and 5.5. Further, the background mosaic is used for region-based segmentation and is done with more accurate results, as shown in Figures 5.4 and 5.5. Compared to traditional methods like SIFT Features, the SURF will have fast and robust features for motion segmentation. Performance measure for the dataset proves Mosaic Based Background Modeling for SURF has fast and accurate registration results, as shown in Figure 5.5.

Figure 5.5 (a) Input frame 220 on MICC dataset, (b) Extracted SURF features, (c) Mosaic background modelling and (d) Foreground segmentation.

5.5   Conclusion

This paper proposes and solves the problem of moving object detection in frame sequences. With PTZ cameras, object segmentation is done using a Background Mosaic Model based on Deep Learning segmentation approaches and then uses the foreground extraction technique. From extracted feature points, the accuracy of segmentation is improved with the help of a SURF feature-based Background Mosaic Model algorithm. The advantages of this method are plenty and, thus, this method is used in a variety of surveillance applications which require tracking and classification. In future work, complex video scenarios with harder conditions will be tested. Trying to detect humans while they are partially or completely occluded is not easy, but using a fast and discriminative descriptor like SURF can solve the problem.

References

1. Abhishek Kumar & Jyotir Moy Chatterjee & Pramod Singh Rathore, 2020. "Smartphone Confrontational Applications and Security Issues," International Journal of Risk and Contingency Management (IJRCM), IGI Global, vol. 9(2), pages 1-18, April.
2. Alessio Ferone and Lucia Maddalena, "Neural Background Subtraction for Pan-Tilt-Zoom Cameras", IEEE Transactions on Systems, Man, and Cybernetics: Systems.
3. Bhargava, N., Bhargava, R., Rathore, P. S., & Kumar, A. (2020). Texture Recognition Using Gabor Filter for Extracting Feature Vectors With the Regression Mining Algorithm. International Journal of Risk and Contingency Management (IJRCM), 9(3), 31-44. doi:10.4018/IJRCM.2020070103.
4. Christopher Broaddus Changjiang Yang Shunguang Wu, Tao Zhao and Manoj Aggarwal, "Robust pan, tilt and zoom estimation for ptz camera by using meta data and/or frame-to-frame correspondences", Control, Automation, Robotics and Vision, 2006.
5. J. Lu, F. L. Li, M. Jiang, "Detection and Tracking of Moving Targets with a Moving Camera", Journal of Harbin Engineering University, 29(8), pp. 831-835, Aug 2008.
6. Jing Li, Quan Pan, Stan. Z. Li, Tao Yang, "Automated Feature Points Management for Video Mosaic Construction", World Academy of Science, Engineering and Technology International Journal of Computer, Information, Systems and Control Engineering Vol:1 No:2, 2007.

7. Kang Xue, Gbolabo Ogunmakin, Yue Liu, Patricio A. Vela, Yongtian Wang, "Ptz Camera-Based Adaptive Panoramic And Multi-Layered Background Model", 2011 18th IEEE International Conference on Image Processing.

8. Kumar, A., Chatterjee, J. M., & Díaz, V. G. (2020). A novel hybrid approach of svm combined with nlp and probabilistic neural network for email phishing. International Journal of Electrical and Computer Engineering, 10(1), 486.

9. Maddalena and A. Petrosino, "A self-organizing approach to background subtraction for visual surveillance applications," IEEE Trans.Image Process., vol. 17, no. 7, pp. 1168–1177, Jul. 2008.

10. N. Bhargava, S. Dayma, A. Kumar and P. Singh, "An approach for classification using simple CART algorithm in WEKA," 2017 11th International Conference on Intelligent Systems and Control (ISCO), Coimbatore, 2017, pp. 212-216, doi: 10.1109/ISCO.2017.7855983.

11. Naveen Kumar, Prakarti Triwedi, Pramod Singh Rathore, "An Adaptive Approach for image adaptive watermarking using Elliptical curve cryptography (ECC)", First International Conference on Information Technology and Knowledge Management pp. 89–92, ISSN 2300-5963 ACSIS, Vol. 14 DOI: 10.15439/2018KM19.

12. Rathore P S, Kumar A "An adaptive method for Edge Preserving Denoising "International Conference on Communication and Electronics Systems, Institute of Electrical and Electronics Engineers", & PPG Institute of Technology. (2017).Proceedings of the 2nd International Conference on Communication and Electronics Systems (ICCES 2017): 19-20, October 2017.

13. Rathore, P.S., Chatterjee, J.M., Kumar, A. et al. Energy-efficient cluster head selection through relay approach for WSN. J Supercomput (2021). https://doi.org/10.1007/s11227-020-03593-4.

14. Singh Rathore, P., Kumar, A., & Gracia-Diaz, V. (2020). A Holistic Methodology for Improved RFID Network Lifetime by Advanced Cluster Head Selection using Dragonfly Algorithm. International Journal Of Interactive Multimedia And Artificial Intelligence, 6 (Regular Issue), 8. http://doi.org/10.9781/ijimai.2020.05.003.

15. Sudipta N. Sinha and Marc Pollefeys, "Pan-tilt-zoom camera calibration and high-resolution mosaic generation", Computer Vision and Image Understanding, vol. 103, pp. 170–183, 2006.

16. X. C. Zhou, D. W. Tu, Y. Chen, Q. J. Zhao, Y. C. Zhang, "Moving Object Detection under Dynamic Background based on Phase-correlation and Differential Multiplication", Chinese Journal of Scientific Instrument, 31(5), pp. 980-983, May 2010.

17. Xinghua Li, Qinglei Chen, Haiyang Chen, "Detection and Tracking of Moving Object Based on PTZ Camera", 2012 IEEE fifth International Conference on Advanced Computational Intelligence (ICACI) October 18-20, 2012 Nanjing, Jiangsu, China.

18. Y. Xiong, K. Turkowski, Creating image-based VR using a self-calibrating fisheye lens, IEEE Proceedings of Computer Vision and Pattern Recognition, Washington, June 1997, pp. 237–243.

19. Z. H. Yang, Y. C. Zeng, "Research on the Compensation Technology of Background Moving", Journal of Beijing Institute of Technology, 20(3), pp. 333-337, June 2000.

6

Prediction of CKD Stage 1 Using Three Different Classifiers

Thamizharasan, K.*, Yamini, P., Shimola, A. and Sudha, S.

Department of Biomedical Engineering, Rajalakshmi Engineering College, Chennai, India

Abstract

Chronic Kidney Disease shows slow and periodical loss of kidney function over a period of time and will develop into permanent kidney failure when left untreated. The proposed work aims at presenting the use of Deep Learning for the prediction of Chronic Kidney Disease. Training has been performed using 16 attributes of about 400 patients. Three Deep Learning techniques (Random Forest, Support Vector Machine, and Naive Bayes Classifier) are helpful in predicting the stages. A comparative analysis of these three classifiers is performed. The results show that a Random Forest Classifier has a higher accuracy rate of about 96.25% and an error rate of 0.03. Prediction of stage 1 is usually challenging, but a Random Forest Classifier has helped in predicting stage 1 of the disease with an accuracy of 75%, making it the most effective classifier. This classification would be helpful for prior forecasting of the disease and avoids future health implications for the patient. It would therefore play a major role in reducing the mortality rate of patients.

Keywords: Chronic kidney disease, deep learning, support vector machine, random forest, naive bayes classifier

6.1 Introduction

Chronic Kidney Disease (CKD) is a condition marked by a gradual loss of kidney function over a period. CKD was declared as the 9th leading cause of death by the National Kidney Foundation in 2019 [1]. The rise in population and unhealthy habitual behavior resulted in the development of Kidney

---

Corresponding author: thamizharasan.k.2016.bme@rajalakshmi.edu.in

Pramod Singh Rathore, Vishal Dutt, Rashmi Agrawal, Satya Murthy Sasubilli, and Srinivasa Rao Swarna (eds.) Deep Learning Approaches to Cloud Security, (75–92) © 2022 Scrivener Publishing LLC

Diseases. Especially in developing countries, the expected shoot up in the number of Kidney Disease cases is predicted to have a profound effect on socio-economic conditions. Kidney Disease is considered a medical challenge in the 21st century. Even though there are a large number of pandemic kidney problems, CKD is widely accepted as the rising menace of this epoch. Being a longstanding disease of the century, CKD is classified into five stages based on the patient's Glomerular Filtration Rate (GFR) and Serum Creatinine levels. Of these the most minacious is the fifth stage of CKD where the patient reaches a GFR value less than 15 ml/min per 1.73 m^2 [2].

Recent landmark studies claim that the fifth stage of CKD eventually ends up in End-Stage Renal Disease leading to ultimate renal dysfunction. Early detection of the CKD stages is the only way to hold the fort unless the condition is inexorable. Though a lot of researchers worked their way through to predict the stages of CKD, prediction of the first stage of CKD is still unforeseeable. Since the GFR value associated with the first stage of this chronic disease is 90 ml/min per 1.73 m^2 that is close to the normal value of GFR in a healthy person, the prediction of this stage has been a formidable challenge [3].

Numerous analysts and researchers are working monotonously for a long time to predict CKD stages. Deep Learning algorithms have been used to predict the presence of CKD in patients. In this work, they used six different Deep Learning algorithms: Naive Bayes, Sequential Minimal Optimization (SMO), Simple Logistic (SLG), Radial Basis-Function (RBF), Multilayer Perceptron Classifier (MLPC), and Random Forest (RF) Classifiers. The results acquired show that the RF Classifier beats different classifiers in terms of area under the ROC curve (AUC), precision, and MCC with a value of 1.0.

The other approach used in the other studies proposed is a versatile neuro-fuzzy derivation framework (ANFIS) used to predict the probability of renal failure due to CKD. The GFR values were predicted by a Takagi-Sugeno type ANFIS model. The model precisely predicted the GFR with an accuracy of 95%, but there were fluctuations in the result when the key attributes are updated [4].

A classification model was developed using a Deep Learning algorithm and the primary aim of this study is to predict whether the patient is at the risk of CKD or not. A dataset with 24-fields having numerical and nominal parameters was used. The numerical field included parameters like age, blood pressure, gender, etc. The nominal field included diabetes mellitus, hypertension, etc. The process used six types of classifying algorithms including a support vector machine using RBF kernel, a support vector machine using linear kernel, an adaptive boosting (AdaBoost), a random forest classifier, decision trees, and logistic regression. Out of these, the SVM with linear kernel gave the best outcome, with an accuracy of 98%, and obtained a purity score of 0.62 after clustering the entire dataset. This work developed a model which accurately predicts the presence of CKD in patients [5].

Though the SVM classifier gave 98% accuracy, further studies proved better results in the prediction of CKD. This study used data from the UCI Deep Learning Repository to validate the different classifiers of the Deep Learning Algorithm. This data is then analyzed using different classifiers including a Support Vector Machine (SVM), Random Forest (RF), k-Nearest Neighbor (k-NN), Artificial Neural Network (ANN), and c4.5 Decision Tree. An open source WEKA (Waikato Environment for Knowledge Analysis) environment is used in this work. In comparison to all the other classifiers, the RF gave the highest precision and F-measure value of 1 and an accuracy of 100%, which proves it as a suitable classifier for predicting the presence of CKD [6].

A data mining algorithm was used for predictive analysis of the advancement of CKD. This study used about 24-prescient parameters with rule-based classifiers and brought about a framework that would foresee CKD and not CKD conditions. The analysis infers that rule-based classifier Decision Table-Naïve Bayes (DTNB) study results are the most accurate area under the ROC curve (AUC) of 99.9% and a least false positive (FP) rate of 1.1%. The discoveries of this relative examination on rule-based studies could be utilized to assist in the forecast of CKD.

Studies on the progression of CKD also used a data mining algorithm with a temporal abstraction (TA) technique. The temporal abstraction is mainly used to study the vital parameters of patients [7]. They used several data mining techniques including a support vector machine, adaptive boosting, c4.5, and a classification and regression tree (CART). The combination of CART and AdaBoost model yielded the most precise value among the classifiers with an accuracy of 66.2%, sensitivity of 62%, AUC of 71.5%, and specificity of 70.4%. All these findings made their model support in the appropriate diagnosis of CKD in patients. Deep Learning algorithms were used to predict the presence of CKD in patients. This work used a support vector machine and decision tree classifiers for processing the data. The project used datasets from a UCI repository and utilized 14 important attributes for the analysis. The outcomes were quite precise. The decision tree algorithm produced an accuracy of 91.75% and the support vector machine algorithm produced an accuracy of 96.75%. Though the SVM produced a more accurate result, the decision tree model needs a shorter time period. The limitation of this model is that the data strength is weak and had many missing attributes [8].

However, works on pre-emptive diagnosis of the disease would promptly diminish the occurrence and costs required for treatment of CKD. This is done by precisely using a data mining algorithm with its classifiers and AI methods. Test results show that ANN, SVM, and Naïve Bayes accomplished an accuracy of 98.0%, while k-NN accomplished a precision of 93.9% [9].

Studies show that the prediction of early detection CKD would negligently lessen the mortality rate. Using Deep Learning algorithms like support vector machine (SVM), J48, and Multilayer Perceptron (MLP), this work mainly focused around finding the best classification algorithm depending on various parameters like accuracy and root mean square using a 2x2 confusion matrix. The results show that MLP provides a precision accuracy of 99.75%. This work could be used to predict the presence of CKD in patients but does not help in predicting the early stages of CKD [10].

On an estimated analysis, over two million patients worldwide are affected by End-Stage Renal Disease characterized by a Glomerular Filtration Rate value less than 15 ml/min per 1.73 m². CKD is a burden not only due to renal replacement therapy (RRT) demands, but also for the health of the people. In this paper we have made an attempt to predict the first stages of CKD with the help of a Deep Learning Algorithm. Since the effect of CKD on global health is prodigious, it is essential to build a system that would ease the prediction of early stages of the disease. In this work, we predicted the stages of CKD by using a Deep Learning algorithm with 16-predictive attributes [11]. The classifiers used are a Random Forest Classifier, Support Vector Machine Classifier, and Naive Bayes Classifier. Each classifier was trained and tested to get the percentage of accuracy using a predicted value. Out of the three classifiers we used, the Random Forest Classifier worked best when compared to the results predicted by a Support Vector Machine and Naive Bayes Classifier. With the help of a confusion matrix, we successfully predicted the stages of CKD using a Random Forest Classifier. The overall accuracy of RF for predicting the stages of CKD is 96.25% and the testing error rate is 0.03%. This work helped us classify the patients relying upon the stages of CKD. We believe the early detection of the phases of kidney disease will surely help patients avoid falling into ESRD and likewise lessens mortality trouble due to kidney maladies [12].

6.2   Materials and Methods

The staging of Chronic Kidney Disease can be predicted using Deep Learning. A sequence of steps which is made to learn by the computer, eventually, is called Deep Learning (ML). These algorithms build a model based on the input data which we have fed into the system; this kind of data is called training data. Deep Learning approaches may be broadly classified under three groups as supervised learning, unsupervised learning, and reinforcement learning based on the type of input which is given to the system [13].

The classification of CKD has been done using three proposed classifiers, namely, Random Forest, Naive Bayes, and Support Vector Machine. The proposed workflow for the prediction of each stage is shown in Figure 6.1. The steps which come under the workflow include acquirement of datasets, preprocessing of data, and training and testing the obtained datas. These steps are further explained below.

A. Acquirement of Dataset
The dataset we have used is taken from the KAGGLE Deep Learning Repository. This dataset contains data for 400 patients with 16 attributes, namely, Age, Glucose, Blood Pressure, Specific Gravity, Albumin, Urea, Creatinine, Sodium, Potassium, Phosphate, Hemoglobin, Packed

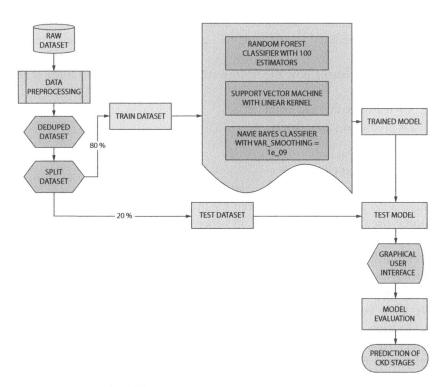

Figure 6.1 Proposed workflow.

Cell Volume, Hypertension, Diabetes Mellitus, Coronary Heart Disease, Edema, and Glomerular Filtrations Rate. Among these 16 attributes, 4 of the attributes show higher correlation when compared to others, namely, Hypertension, Urea, Diabetes Mellitus, and Creatinine. Among 400 patients, 250 are classified as patients with CKD and the remaining 150 as normal patients without CKD [14].

B. Data Preprocessing

The dataset contains some missing values or null values and certain attributes like Hypertension, Coronary Artery Disease (CAD), Diabetes Mellitus, Edema, and Anemia have the values listed as YES or NO. These missing values are replaced by average value of the entire attribute and the values which had YES [15].

Dataset Training

The data from the dataset is trained by using three different classifiers. The training dataset is taken from the cleaned dataset; out of 400 patients 80% is used for training the model, i.e., 340 patients' data. This training data is further split into dependent class or feature columns and predicted class.

feature columns = [„glucose", „age", „bp", „gravity",
„albumin", „urea", „creatinine", „sod", „pot", „hemo", „pcv",
„htn", „D.mellitus", „cad", „edema", „anemia", „gfr"] predicted_class =
[„ckd_stage"]

This model learns how the feature column value would affect the stage of CKD

X = feature columns, Train
Y= predicted_class, Train

C. Testing of Datasets

Random Forest Classifier

This classifier is one of the compatible and easier algorithms to use. Also, there is no proper tuning required for the classifier. It is one of the commonly used algorithms among all other algorithms [16]. This could be used for classification and regression. This kind of classification method uses multiple numbers of decision trees during the training phase for producing the desired output and that becomes the mode or mean of each and every tree. They do not suffer from the problem of over fitting.

Decision Tree

A Decision Tree is the fundamental unit of this type of classifier. It comes under supervised learning. By means of decision trees, it classifies values better. Every tree has a node which represents an attribute and every leaf node represents a class. In random forest pseudo code, the first step includes placing the best fit property or attribute at the root. The second step involves dividing the training dataset in subsets of data and these subsets should contain the same value for a particular feature. These two steps need to be repeated until there are leaf nodes contained in every tree's branch. Always, the attribute of the root is compared with that of the records. Based on this, we select the branch with the corresponding value and move to the other node. This process continues until we find a leaf node with a predicted value [17].

It is always necessary to have a correct ratio of samples to feature numbers. It is essential to use the reduction of dimensions which gives better discriminative features. It is important to visualize the tree as an export function. It is also important to increase the depth of trees step by step for preventing the problem of over fitting. Whenever the sample size varies, float number could be used as a percentage. For a fewer number of classes, classification could be taken for a leaf value of one [18].

The algorithm of this classifier works simply under four steps.

Step 1: Selection of arbitrary samples from the obtained dataset

Step 2: Decision tree needs to be built for every sample and every predicted result needs to be obtained from every decision tree

Step 3: Voting needs to be performed for the predicted result

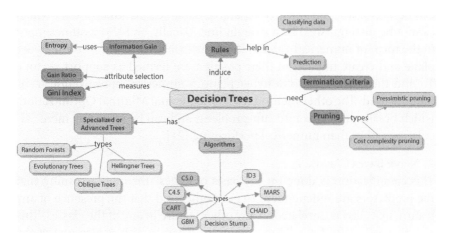

Figure 6.2 Working of decision tree.

Step 4: The predicted result with the majority number of votes is selected as the final prediction. This is explained in the decision trees shown in Figure 6.2. The magnified view of decision trees is explained in Figure 6.2.

1) Support Vector Machine Classifier

Support Vector Machines (SVM) come under supervised learning. Similar to Random Forest classifiers, they can be used for classification and regression (Singh, N., Singh, P. (2016)). An SVM classifier is best represented by coordinates in space that are plotted in such a way that the examples belonging to different categories are divided by a line which must be as wide as it could be placed. Newer categories are placed into the same place as it was earlier and they are predicted based on the gap side they belong to [19].

A Support Vector Machine (SVM) plots a hyper plane usually in an n-dimensional space and can be used for classification or regression. On the hyper plane, whichever point has the maximum distance to the nearest point in any of the classes is selected. However, it is always easier to have a linear plane for the separation of two classes. This hyper plane is not manually inserted. In this case, kernels are usually used. An SVM kernel usually takes a low dimensional space and in turn converts into a higher dimensional space. This algorithm works well under a perfect separation line. It also turns out to be highly effective for higher dimensional spaces and also this classifier is memory-efficient [20].

An SVM is usually classified as one of two types: linear or non-linear. A linear SVM is used for linearly separable data because it classifies two different classes by means of a single straight line, whereas a non-linear SVM can be useful for non-linearly separable data, i.e., the kind of data which cannot be distinguished by a straight line. Usually, an SVM assumes input in the form of numerical values. The points which are nearer to the hyper plane and create impact on their position are termed as support vectors. To find the coefficients of a hyper plane, a numerical optimization technique is used. The other method used is Sequential Minimal Optimization, which breaks problems into sub-problems and can be solved by mere calculating rather than numerical performing [21].

2) Naive Bayes Classifier

This classification is done on the basis of Bayes" theorem assuming that the predictors are independent to each other and that the presence of any feature in a class is unrelated to any other feature in any of the classes. This classifier is helpful for a large number of datasets. This is also one of the perfect classification methods used because it is simpler [22]. Usually, the dataset is classified into a feature matrix and response vector where the feature matrix has all vectors in which every vector contains values of dependent features and the response vector contains the class variable value for every row of the feature matrix.

This classifier is based on the assumption that every feature contributes independently and equally to the outcome. Some of the popular Naive Bayes Classifiers include Multinomial and Bernoulli Classifiers. In case of multinomial classifiers, frequencies are represented by feature vectors that are generated by multinomial distribution, whereas in Bernoulli, features are usually independent binary variables. Inputs in a Naive Bayes algorithm usually undergo of the following steps:

Step 1: Convert the entire data set into a table of frequencies.
Step 2: Create a likelihood table on the basis of certain probabilities.
Step 3: Use the Bayesian Equation for calculating the posterior probability for each and every class; the one with the highest posterior probability is taken to be the predicted outcome.

Navie Bayes Classifiers usually require a small number of training data to find out the required parameters and they are extremely fast. They also eliminate certain dimensional problems and fewer amounts of training data is enough for better prediction. This runs well when categorized inputs are given rather than numerical variables. In case of numerical variables, bell curves are usually used [23].

D. Classification of Data
First, the classifiers are imported from a sklearn module and are assigned a particular variable name. Then, required parameters are fed to the classifiers for setting them up in tandem with our machine model. Training data is then fitted to the classifiers for training the model. Once the training is complete, the testing data is fed to the machine model for prediction of test dataset.

E. Model Evaluation
Predicted values are collected and compared to the original values from the test dataset from sklearn using a Metrics Module Import Confusion Matrix, Accuracy Score, Classification Report, Cohen Kappa Score, and Mean Squared Error. The Confusion Matrix provides the correlation between true value and predicted value. A Classification Report provides parameters such as an F1 score, Precision, and Recall [24].

1) **Accuracy** indicates how correct the measurement is. It tells how close a measurement is to a particular value. It shows the difference between the result and the true value.

$$Accuracy = TP+TN/(TP+TN+FP+FN)$$

2) **Precision** tells how far two measurements are related with each other. Precision is usually calculated by using a standard value as a reference value.

$$Precision = TP/TP+FP$$

3) **Recall** is usually used as a method for relevance. It is usually a fraction of successfully obtained relevant datas.

$$Recall = TP/TP+FN$$

4) The **F1 Score** is one of the methods performed for determining accuracy. It is calculated as the average of precision and recall. F1 score is calculated using

$$F1\ Score = 2^*(Recall\ ^*\ Precision) / (Recall + Precision)$$

5) The **Cohen Kappa Score** is a measurement of how often the instances classified by means of Deep Learning Classifier match the data, which is usually maintained as a standard reference.

$$Kappa = (OA - AC) / (1\text{-}AC)$$

where

OA – Observed Agreement
AC - Agreement of Change

6) **Root Mean Square Error** is the square root of the Mean Square Error. It is taken as the square of the difference between an estimated value and that of the parameter. It is a frequently used measure of the differences between values predicted by a model and the values actually observed [25].

6.3   Results and Discussion

A. Graphical User Interface
Graphical User Interface makes the operation more intuitive and easier for the user. In order to view this model from the application perspective,

a user interface has been developed. This user interface helps the user to enter in patient data and find out the exact stage the patient belongs to. The layout is designed by a Pyqt5 module and the following steps aid in the access of a developed interface.

Step 1: Run the python file in the command prompt
Step 2: Enter the patient parameters
Step 3: Stage of CKD predicted will be displayed

The values entered are temporarily stored in CSV file and it is replaced once the next data is entered.

The Graphical User Interface displaying the CKD stage of the patient in three different classifiers is displayed in Figure 6.3.

The mean and standard deviation of urea and glucose for the training and testing dataset has been visualized in Figures 6.4 and 6.5. A higher glucose level eventually leads to diabetes mellitus. Since diabetes mellitus has a correlation factor of 0.57, this has a major contribution in the onset of disease.

The mean and standard deviation values of creatinine for the train and test dataset have been displayed in Figure 6.6. The correlation factor for creatinine is 0.49, which shows that it majorly contributes to the progression

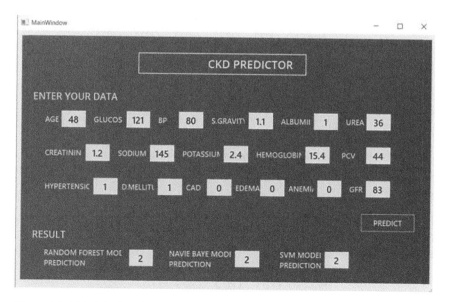

Figure 6.3 Main Window displaying user interface.

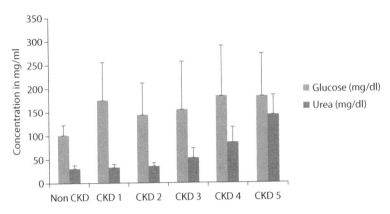

Figure 6.4 Mean & standard deviation for urea and glucose (training data).

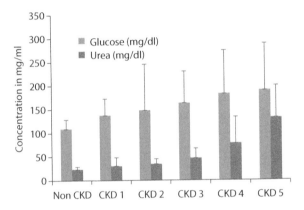

Figure 6.5 Mean & standard deviation for urea and glucose (test data).

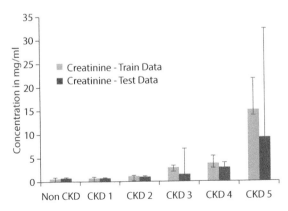

Figure 6.6 Mean & standard deviation for creatinine (training & test data).

of the disease. People with higher levels of creatinine may belong to later stages of disease [26].

B. Classification of Test Data
The test data is classified by each of the three classifiers as follows. Certain parameters such as Accuracy, F1-Score, Precision, Recall, Kappa, and Root Mean Square Error are shown in Table 6.1 [27].

The comparative analysis of accuracy and error of three different classifiers, as displayed in Figure 6.7 shows, that the Random Forest classifier showed the highest accuracy (96%) and a least Root Mean Square Error of 0.03 [28]. It is seen that the Random Forest classifier performs better than an SVM with an accuracy of 88% and Naive Bayes classifier with 75% accuracy, as seen in Figure 6.7.

The following report gives evaluation parameters including Accuracy, F1 Score, and Precision to each and every stage of CKD that are present in the test dataset.

Table 6.1 Classified results of classifiers.

Parameters	Random forest classifier	SVM classifier	Naive Bayes classifier
Total Number of Instances	80	80	80
Correctly Classified Instances	77	71	60
Incorrectly Classified Instances	3	9	20
Accuracy (%)	96	88	75
F1-Score	0.96	0.89	0.74
Precision	0.97	0.90	77
Recall	0.96	0.89	75
Kappa	0.95	0.85	0.67
Root Mean Square Error	0.03	0.125	0.3

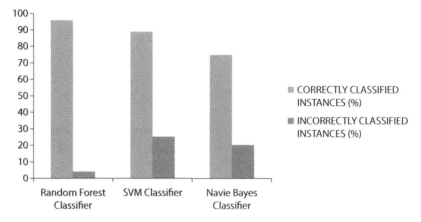

Figure 6.7 Comparative analysis of accuracy & error of different classifiers classification report.

Table 6.2 Random forest classification report.

Stage	Accuracy (%)	Precision	F1-score
0	96	0.96	0.96
1	75	0.75	0.75
2	83	0.83	0.91
3	88	0.88	0.80
4	69	0.69	0.78
5	100	1.00	0.96

Table 6.3 Naïve bayes classification report.

Stage	Accuracy (%)	Precision	F1-score
0	100	0.82	0.90
1	0	0.00	0.00
2	40	0.67	0.50
3	68	0.93	0.79
4	90	0.50	0.64
5	57	0.89	0.70

Table 6.4 Support vector machine classification report.

Stage	Accuracy (%)	Precision	F1-score
0	100	0.97	0.98
1	75	1.00	0.86
2	100	0.83	0.91
3	89	1.00	0.94
4	100	0.91	0.95
5	100	1.00	1.00

The experimental results infer that classifying Stage 1 is usually tedious for classifiers. Random Forest and SVM classifiers both have the same accuracy of 75% for Stage 1, which is seen as in Figure 6.7 [29]. The results show that the Random Forest classifier shown in Table 6.2 showed better accuracy for other stages when compared to SVM and the Naive Bayes classifier showed the least accuracy for all stages shown in Table 6.3 and Table 6.4 [30].

6.4 Conclusions and Future Scope

In this study, we have employed three different Deep Learning methods for the prediction of Chronic Kidney Disease [31]. From the results obtained, it is seen that the Random Forest Classifier shows the maximum accuracy at 96.25% and an error rate of 0.03. Prediction of Stage 1 is more precisely done by the Random Forest Classifier with an accuracy of 75% when compared to the other classifiers. In the future, a much more accurate system could be devised for predicting the progression of the disease with a comparatively lesser number of attributes [32].

References

1. Alassaf, Reem A. *et al.* (2018) - "Preemptive Diagnosis of Chronic Kidney Disease Using Deep learning Techniques."- International Conference on Innovations in Information Technology (IIT), pp. 99-104.
2. Bhavani R. *et al.* (2019) – "Vision-Based Skin Disease Identification Using Deep Learning" - International Journal of Engineering and Advanced Technology (IJEAT), Vol. 8, no. 6, pp. 2249 –8958.

3. Singh Rathore, P., Kumar, A., & Gracia-Diaz, V. (2020). A Holistic Methodology for Improved RFID Network Lifetime by Advanced Cluster Head Selection using Dragonfly Algorithm. International Journal Of Interactive Multimedia And Artificial Intelligence, 6 (Regular Issue), 8. *http://doi.org/10.9781/ijimai.2020.05.003*

4. Eyck, Jo Van *et al.* (2016)-"Prediction of Chronic Kidney Disease Using Random Forest Deep learning.

5. Algorithm."https://blog.transonic.com/hemodialysis/morbidity-and-mortality-in-patients-with-CKD

6. https://www.davita.com/education/kidney-disease/stages/stage-1-of-chronic-kidney-disease.

7. Naveen Kumar, Prakarti Triwedi, Pramod Singh Rathore, "An Adaptive Approach for image adaptive watermarking using Elliptical curve cryptography (ECC)", First International Conference on Information Technology and Knowledge Management pp. 89–92, ISSN 2300-5963 ACSIS, Vol. 14 DOI: 10.15439/2018KM19

8. Rathore, P.S., Chatterjee, J.M., Kumar, A. *et al.* Energy-efficient cluster head selection through relay approach for WSN. J Supercomput (2021). https://doi.org/10.1007/s11227-020-03593-4

9. https://www.kidney.org/news/newsroom/factsheets/KidneyDiseaseBasics

10. https://www.ncbi.nlm.nih.gov /pmc/articles/PMC6308749/

11. Abhishek Kumar & Jyotir Moy Chatterjee & Pramod Singh Rathore, 2020. "Smartphone Confrontational Applications and Security Issues," International Journal of Risk and Contingency Management (IJRCM), IGI Global, vol. 9(2), pages 1-18, April.

12. https://www.ncbi.nlm.nih.gov/pmc/articles/PMC4089693/

13. Bhargava, N., Bhargava, R., Rathore, P. S., & Kumar, A. (2020). Texture Recognition Using Gabor Filter for Extracting Feature Vectors With the Regression Mining Algorithm. International Journal of Risk and Contingency Management (IJRCM), 9(3), 31-44. doi:10.4018/IJRCM.2020070103

14. Jadhav, Sayali & Chandran, Priya & Vijaykumar, Suhasini (2019). - "Comparative Study of Chronic Kidney Disease Prediction using Deep learning Techniques."- International Journal of Computer Sciences and Engineering, vol. 7, pp. 501-506.

15. N. Bhargava, S. Dayma, A. Kumar and P. Singh, "An approach for classification using simple CART algorithm in WEKA," *2017 11th International Conference on Intelligent Systems and Control (ISCO)*, Coimbatore, 2017, pp. 212-216, doi: 10.1109/ISCO.2017.7855983.

16. Jeni, László A *et al.* (2013) - "Facing Imbalanced Data Recommendations for the Use of Performance Metrics." - International Conference on Affective Computing and Intelligent Interaction and workshops : [proceedings]. ACII (Conference), Vol. 2013, pp.245-251.

17. Jo, Taeho *et al.* (2019). – "Deep Learning in Alzheimers Disease: Diagnostic Classification and Prognostic Prediction Using Neuroimaging Data." - Frontiers in aging neuroscience, Vol. 11, no. 220.
18. John Cho, Manoj Reddy. (2016)- "Detecting Chronic Kidney Disease Using Deep learning" - Hamad bin Khalifa University Press, Vol.2016 No. 1, pp.1-8.
19. Mantzaris D. *et al.* (2007) – "Abdominal Pain Estimation in Childhood based on Artificial Neural Network Classification" – CEUR Workshop Proceedings, Vol. 284.
20. Mehtaj Banu H (2019) –"Liver Disease Prediction using Machine- Learning Algorithms" - IJEAT, Vol. 8, no. 6, pp. 22.
21. Moore, P J *et al.* (2019) - "Random forest prediction of Alzheimer disease using pairwise selection from time series data." PloS one, Vol. 14, no. 2.
22. Mythili T *et al.* (2013) – "A Heart Disease Prediction Model using SVM-Decision Trees-Logistic Regression (SDL)" – International Journal of Computer Applications, Vol. 68, no. 16.
23. N. Bhargava, A. Kumar Sharma, A. Kumar and P. S. Rathoe, "An adaptive method for edge preserving denoising," 2017 2nd International Conference on Communication and Electronics Systems (ICCES), Coimbatore, 2017, pp. 600-604, doi: 10.1109/CESYS.2017.8321149.
24. N. Bhargava, S. Sharma, R. Purohit and P. S. Rathore, "Prediction of recurrence cancer using J48 algorithm," 2017 2nd International Conference on Communication and *Electronics Systems (ICCES)*, Coimbatore, 2017, pp. 386-390, doi: 10.1109/CESYS.2017.8321306.
25. Norouzi, Jamshid *et al.* (2016) - "Predicting Renal Failure Progression in Chronic Kidney Disease Using Integrated Intelligent Fuzzy Expert System." - Computational and mathematical methods in medicine, Vol. 2016.
26. Powers, D.M.W., (2011) – "Evaluation: from Precision, Recall and F-measure to ROC, Informedness, Markedness and Correlation." -Journal of Deep learning Technologies, Vol. 2, no. 1, pp. 37-63.
27. Siddheswar Tekale, Pranjal Shingavi, Ankit Chatorikar (2018) – "Prediction of Chronic Kidney Disease Using Deep learning Algorithm"- International Journal of Advanced Research in Computer and Communication Engineering Vol. 7, Issue 10.
28. Kumar Sahwal, ..., Kishore, ..., Singh Rathore, ..., & Moy Chatterjee, ... (2018). An Advance Approach of Looping Technique for Image Encryption Using in Commuted CONCEPT OF ECC. International Journal Of Recent Advances In Signal & Image Processing, 2(1).
29. Singh N, Singh P. (2016). – "Rule Based Approach for Prediction of Chronic Kidney Disease: A Comparative Study." - Biomed Pharmacol J, Vol. 10, no. 2.
30. Subasi A., Alickovic E., Kevric J. (2017) – "Diagnosis of Chronic Kidney Disease by Using Random Forest" In: Badnjevic A. (eds) CMBEBIH 2017. IFMBE Proceedings, Vol. 6, pp. 589-594.

31. Vijayarani S *et al.* (2015) – "Liver Disease Prediction using SVM and Naïve Bayes Algorithms"- International Journal of Science, Engineering and Technology Research (IJSETR) Vol. 4, no. 4.

32. Kumar, A., Chatterjee, J. M., & Díaz, V. G. (2020). A novel hybrid approach of svm combined with nlp and probabilistic neural network for email phishing. International Journal of Electrical and Computer Engineering, 10(1), 486.

7

Classification of MRI Images to Aid in Diagnosis of Neurological Disorder Using SVM

Phavithra Selvaraj, Sruthi, M.S.*, Sridaran, M. and Dr. Jobin Christ M.C.

Department of Biomedical Engineering, Rajalakshmi Engineering College, Chennai, India

Abstract

This chapter provides a cognizance into the implementation of a support vector machine algorithm of Deep Learning to diagnose neurological conditions. With the advancement of emerging technologies, the means of diagnosing neurological conditions is substantially more complex than it used to be. Procedures and diagnostic tests are tools that help doctors identify a neurological illness or other medical condition. Precise identification of neurological pathologies can be done by adhering to an autopsy after the person's passing. This paper is a novel implementation of an SVM algorithm to classify and diagnose various neurological pathologies using MRI images.

Keywords: SVM, neurological disorders, MRI, deep learning

7.1 Introduction

Mental wellbeing plays a substantial part in a person's health. With the number of neurological diseases increasing lately, the treatment gap for patients grows larger as the number of neurologists present is significantly less compared to the growth in the number of patients. Due to the repetition of many common symptoms between the many conditions in

---

Corresponding author: sruthi.ms.2016.bme@rajalakshmi.edu.in

Pramod Singh Rathore, Vishal Dutt, Rashmi Agrawal, Satya Murthy Sasubilli, and Srinivasa Rao Swarna (eds.) *Deep Learning Approaches to Cloud Security*, (93–108) © 2022 Scrivener Publishing LLC

various disorders, evaluating damage to the nervous system has been a complex task.

Medical Imaging technology has had various ground-breaking advancements in diagnosing neural conditions. Numerous studies have been carried out by scientists to improve methods in order to provide elastic diagnostic information. Medical Imaging of the brain incorporates various types of imaging techniques used to diagnose stroke, blood vessel malformations, tumors, trauma, brain growth abnormalities, and hemorrhage in the brain. Magnetic Resonance Imaging uses radio waves generated by a programmed computer to generate detailed images of body tissues [1].

An MRI can be used for the diagnosis of several conditions some of which are mentioned below:

- Stroke
- Traumatic brain injury
- Brain and spinal cord tumors
- Inflammation
- Infection
- Vascular irregularities
- Disorders like Multiple Sclerosis
- Brain damage associated with epilepsy
- Abnormally developed brain regions
- Some neurodegenerative disorders

Papers about the classification of brain MRI images to identify conditions were studied and analyzed in order to arrive at the most optimal method for carrying out classification of the Brain MRI Images for diagnosing several neural conditions. Research studies on the evaluation of Deep Learning algorithms were studied and papers based on the structural features in MRI that are used for diagnosis of neurological conditions were also analyzed. One of the papers concluded that the selection of appropriate feature type is of the utmost importance when it comes to the efficiency of classification. The evaluation of different classifiers proved that most classifiers produced results with more or less the same accuracy when an adequate feature type is selected.

Comparative analysis of different conditions and the correlation of MRI images with respect to them were also performed and the conclusions closely studied. One such study includes the classification of autism through Brain MRI Analysis aiming to develop an Automated Cognitive System to classify ASD from healthy controls and to predict the neurotransmitter pattern in ASD patients which enables them to identify the

Autism affected region easily. Similarly, MRI findings in Schizophrenia which described the structural changes in Schizophrenia affected patients' brain using MRI were studied too. Progressive changes in the frontal lobe, basal ganglia changes, and amygdala hippocampal volume reduction was reported [2].

The basal ganglia changes were more related to the type and duration of neuroleptic treatment. Similarly, structural findings of MRI in Parkinson's disease were also studied.

7.2   Methodology

Computerized medical image analysis and computer-aided diagnosis has progressed rapidly, promoting many imaging techniques to find applications in medical image processing. Several image processing operations should be applied on brain images before they can be used for the diagnosis of neurological pathologies. The identification, segmentation, and detection of an infected area in brain MRI images are tedious processes. The Magnetic Resonance Imaging technique provides a clear image on the neural architecture of the brain [3].

An MRI contains many imaging modalities to scan and capture the internal structure of the human brain. The flowchart of the methodology used in this work is shown in Figure 7.1.

7.2.1   Data Acquisition

The MR images used in this work are obtained from four different databases. Autism MR images were obtained from an online repository called the LONI Image and Data Archive, which is managed by the University of Southern California. The IDA is a collection of data from different studies. Among these, the Autism Brain Imaging Data Exchange (ABIDE) consists of MR images of Autism patients. It consists of images with dimensions 176x256x256 and voxel size 1.05x1.05x1.05 [4].

The Parkinson's MR images and Schizophrenia MR images are obtained from Open Neuro and Open fMRI, which are online databases consisting of images from different research studies. The Parkinson's image dimensions are 192x256x256 and the voxel size is 1x1x1. The MR images of patients with Schizophrenia have dimensions of 160x192x192 and voxel size -1x1x1. The MR images of healthy individuals (control images) are obtained from both the above databases.

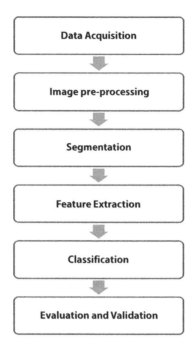

Figure 7.1 Flowchart

7.2.2   Image Preprocessing

The most vital step in any image-based analysis is the preprocessing which performs operations on the image to enhance the image so that the images can be interpreted in the right way without any errors in the upcoming steps. The basic steps to be carried out in MR image processing are Filtration, Contrast Enhancement, and Skull Stripping [5].

a) Median Filtering
Median filtering is a non-linear filtering technique which is used to remove noise from the input by smoothening the parts of the image except the edges. Median filtering mostly focuses on the preservation of edges. Each pixel in the image is compared with its neighboring pixels. The pixel is then replaced with the median of the surrounding pixels. The median filter removes salt and pepper noise with greater efficiency. Since MR images are prone to salt and pepper noise, the median filtering technique is used to eliminate unwanted noise parameters [6].

b) Contrast Enhancement
Adaptive Histogram Equalization (AHE) is a process to enhance the contrast of the image. This method computes several histograms for different

sections of the image and these are used to distribute the lightness value all over the image while the normal histogram technique used in the histogram is derived from the whole image to transform all the pixels in the image. The adaptive histogram equalization technique is widely used because it enhances the local contrast and edge definition in every section of the image [7]. The ordinary histogram equalization technique suitable for images does not have large variations in the pixel value throughout the image. Adaptive Histogram Equalization (AHE) can sometimes over-amplify the homogenous regions of the image since the regions are highly concentrated. To prevent this, a varied technique called Contrast Limited Adaptive Histogram Equalization (CLAHE) is used. The clipping of the histogram is done at a predefined value before the computation of the CDF in CLAHE, thereby limiting the slope of CDF and the transformation function. This predefined value where the clipping of histogram takes place is called the clip-limit and depends on the histogram normalization and the size of the neighbourhood region. Common values that limit the resulting amplification are between 3 and 4.

c) Skull Stripping
Magnetic Resonance Imaging is a high resolution imaging technique which also contains non-brain tissues like skull, skin, fat, muscles, the neck, and eyeballs. These non-brain tissues are considered as obstacles in image processing and analysis procedures like automatic segmentation. Therefore, a preliminary process to isolate brain tissues from extra-cranial or non-brain tissues is always required before any quantitative morphometric studies of MRI brain images. This process is called Skull Stripping.

The Skull Stripping in this work is done by initially converting the image into a binary image with the help of a threshold and eliminating the connected pixels which are less than a particular number, producing a binary mask for the image. Then, this mask is used to isolate the central region of the MR image where the main image is present. After this process, the borders of the masked region are eroded by about 15 to 20 pixels to obtain a Skull Scrapped image. This helps us eliminate the non-brain region of the MR image in a simple way [8].

7.2.3 Segmentation

The process of dividing or partitioning a digital image into multiple divisions or segments is called Image Segmentation. This process is mainly used to separate the region of interest from the rest of the image and to locate the objects and boundaries in an image. Segmentation is a pre-eminent step in image analysis. One of the most successful methods in Image Segmentation is Otsu's Method. Otsu's Method is an automatic threshold selection method which works on the basis of region based segmentation [9].

Otsu's Method is a class variance method which has the ability to select the threshold automatically [10]. This method computes the probability of each intensity level, iterates along every possible threshold value, and creates an evaluation of the entire pixel layout in the image. The desired threshold corresponds to maximum intra-class variance. The main goal of Otsu's segmentation is to reduce the computation of the sum of foreground pixels and background pixels while selecting a fair threshold value which can perform efficient segmentation. Otsu's Method is known for its rare time consumption and is one of the most efficient methods for threshold selection.

Better threshold selection with respect to uniformity and shape measures for real time images is produced by Otsu segmentation. However, a profound search is required to derive the criteria for expansion of within-class variance [11]. This method gives us the total brain region neglecting the other unwanted regions present in the MR image.

7.2.4   Feature Extraction

The process of reducing the dimensions of the image is called Feature Extraction. This reduces a raw set of data into manageable groups for processing. Certain features, parameters, properties, and features are measured from the original set of data to distinguish one input pattern from another. The features extracted are given as classifier inputs [12].

An abnormal brain image will show certain differences in features when compared to a normal brain image. The changes in the features depend upon the type of brain pathology and their cognitive abilities and disabilities. These features are stored into a feature space. In the presence of a small dataset, few instances of abnormalities of a particular neurological disorder will be present in each image. Once the classifier detects an example of the abnormality present in a particular region of the human brain, it learns and produces a model to classify the images as abnormal tissues or corresponding disorder depending on the inputs given.

A crucial step in feature extraction is the selection of features. Good texture features and neighborhood aggregations of the neighborhood intensity data provide better results rather than including individual intensity values [13]. A number of techniques are used for the purpose of feature extraction from MR Images. Some of the common feature extraction techniques are Discrete Wavelet Transform (DWT), Gabor Filters, and a Gray Level Co-Occurrence Matrix. The Gray Level Co-Occurrence Matrix technique is used for this work. The Gray Level Co-Occurrence Matrix (GLCM) outperforms other techniques in terms of the dimension of the feature vectors, thus this technique is considered more appropriate for MRI image classification [14]. GLCM is a statistical technique that extracts texture features from the image. Normal tissues are much different in texture when compared to the abnormal tissues. The texture features which are extracted from GLCM are contrast, correlation, energy, and homogeneity.

Selection of a good set of features also improves efficiency of classification. Additional features were also extracted from this matrix, namely mean, standard deviation, entropy, root mean square, variance, kurtosis, and slowness. In addition to this, the total brain area is estimated and used for classification. When more than two features are used for classification process, it is very difficult to visualize the performance of the classifier. Thus, the features are reduced into two variables that prominently describe all the features derived. This is another process of dimensionality reduction and the Principal Component Analysis technique is used to achieve this. This technique increases interpretability and minimizes information loss at the same time. Thus, visualization can be done with these two prominent variables which are obtained by PCA as a two-dimensional graph [15].

7.2.5 Classification

The process of categorizing the given input by a proper classifier is called classification. The objective of classification is to group items that have similar features into sections. We have a number of classifiers or algorithms to choose from which differ in the way they evaluate the right class for a particular test data. One has to consider the requirements and the type of dataset to choose a suitable algorithm from different classifiers. The k-Nearest Neighbour classifier is a simple, non-parametric algorithm which assigns classes to the test data by considering its Euclidean distance with other points [16].

This algorithm works well if there is a small number of features and the performance of this algorithm is worse in situations where there are many features and only a few are informative.

Naïve Bayes classifiers are a family of classifiers containing algorithms based on Bayes' Theorem in which every pair of features that are being classified are considered to be independent of each other [17]. The extension of Naïve Bayes is the Gaussian Naïve Bayes classifier wherein a Gaussian function is used to estimate the distribution of data. A Gaussian function (or Normal distribution) is the easiest to work with because the estimation of just two parameters, mean and standard deviation from the training data, will help in decision making. In a Syntactic Ambiguity Study, GNB and

SVM outperformed the KNN classifier. The GNB Classifier is inferior in performance with MRI data when compared to a Linear Regression and Linear SVM classifier.

GNB is a useful classifier for procedures that need to be repeated a number of times. Linear classifiers are more appropriate for MRI analysis because there's no problem of overfitting. Linear Discriminant Analysis is a dimensionality reduction technique. This technique reduces the dimensions of the image without much information loss. LDA needs to be used with extreme feature selection, otherwise there is typically not enough data to estimate the covariance matrix reliably [18].

A linear equation fitting relationship between two variables in the observed data is modelled in Linear Regression. LR and SVM tend to be roughly equivalent in terms of performance other than that the first lends itself more naturally to cases where there are more than two classes. A Linear Support Vector Machine Classifier can classify the data efficiently, thus this algorithm is used for this project.

Support Vector Machines are used for this work because they shows greater accuracy in comparison to other classification algorithms on MRI images. A Support Vector Machine is a supervised learning technique that uses a set of labelled training data as input and produces an input-output mapping function as the output. It is a binary classification method that takes features from two classes as input and outputs a model file for classifying new unlabelled/labelled data into one of those two classes. Support Vector Machines are initially two class classifiers that are more systematic to learning linear or non-linear class boundaries and, thus, seem attractive. The SVM classifier used in this work provides the possibility of choosing a suitable kernel function. RBF (Radial Basis Function Kernel) is chosen as a kernel function based on the excellent performance of SVM kernel function reported in many studies. This work utilizes a multi-class SVM classifier in order to classify four classes [19].

7.3   Results and Discussions

7.3.1   Preprocessing

The results obtained from various stages of preprocessing and carrying out the classification were analyzed stagewise and validation of the classification was performed shown in Figure 7.2.

Noise Image   Filtered Image

Filtered Image   Clahe Image

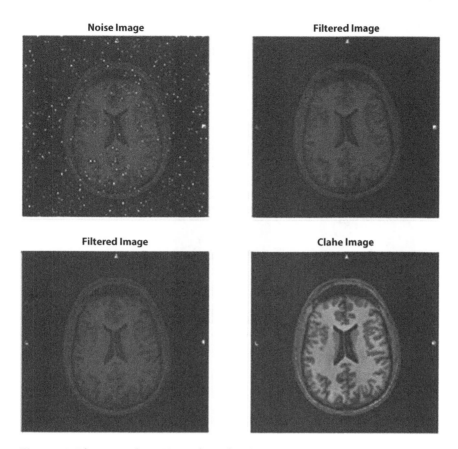

Figure 7.2 Filtration and CLAHE performed in the image.

By applying suitable image processing techniques, significant information and data were extracted from the image. The filtered CLAHE image was obtained by performing Median filtering, after which contrast, enhancement was done using a special form of Adaptive Histogram Equalization. Skull Stripping is also done to produce accurate feature values during feature extraction [20].

Otsu's Segmentation method is used in image processing to perform histogram-based image thresholding or to transform a gray level image into a binary image. This model assumes that the image embraces a bi-modal histogram (for instance, foreground and background pixels) and evaluates the optimum threshold, portioning the two classes so that their intra-class variance is negligible shown in Figure 7.3 and Figure 7.4 [21].

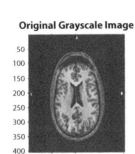

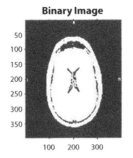

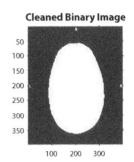

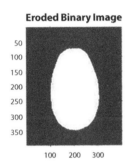

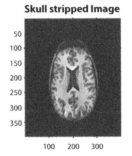

Figure 7.3 Skull Stripping.

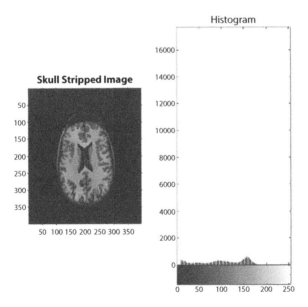

Figure 7.4 Otsu's segmentation.

7.3.2   Classification

The Support Vector Machines used here are more systematic to learning linear or non-linear class boundaries and, thus, seem attractive. Here, we use the MR images of three different disorders (Autism, Parkinson's Disease, and Schizophrenia) and normal MR images to train the classifier and produce a prediction of the input image within these four classes. Therefore, Multi-class Support Vector Machines are used in this case for classification [22].

The results are visualized as a two-dimensional graph plotted between Variable 1 and Variable 2 obtained from the Principal Component Analysis results. Figures 7.5 and 7.6 show the two-dimensional visualization of the hyperplanes obtained for the four classes. The regions are colour indicated with yellow, cyan, red, and green for normal, Schizophrenia, Autism, and Parkinson's Disease classes, respectively [23].

The Support Vector Machine (SVM) classifier is trained after extracting features and labels from the training set. The classifier trains itself and gets an idea and intelligence to classify unknown data from this training data set. After performing the classification, a confusion matrix or error matrix is generated to evaluate the performance of the classifier [24]. Table 7.1 shows the confusion matrix obtained from the results.

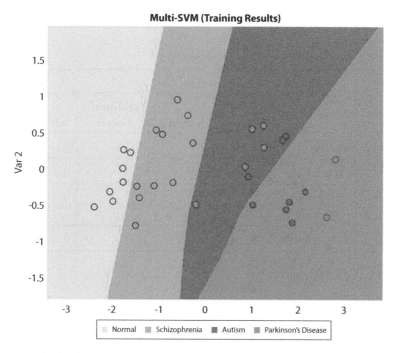

Figure 7.5 Multi-class SVM training results.

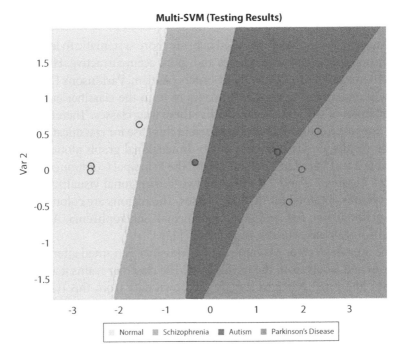

Figure 7.6 Multi-class SVM testing results.

Table 7.1 Confusion matrix – hold out method.

		Predicted class			
		Autism	Normal	Parkinson's disease	Schizophrenia
Actual Class	Autism	1	1	0	0
	Normal	0	3	0	0
	Parkinson's Disease	0	0	3	0
	Schizophrenia	0	0	0	0

7.3.3   Validation

Validation of the efficiency of implemented classification is analyzed by forming a confusion matrix to test for its efficiency. This confusion matrix is formed by using K-Fold cross validation, where the K value was set to 5. In this process, the data set is divided into five different batches randomly and five different experiments on classification are conducted to validate the classifier.

Every batch of data will be used as testing data in each experiment so that the evaluation of the classifier can be appropriate. Table 7.2 shows the confusion matrix obtained by performing K-Fold cross validation [25].

The accuracy, sensitivity, specificity, precision, and F1 score are also evaluated using the above confusion matrix for proper evaluation of the classifier. Table 7.3 gives us the evaluation variables of the classifier [26].

From the below table, it is clear that this algorithm works well with Normal and Schizophrenia classes and the features extracted from Autism and Parkinson's disease should be improved in order to attain higher accuracy. The overall accuracy attained by this algorithm is 85% and the overall F1 score obtained is 0.773 [27].

Table 7.2 Confusion matrix – K-fold method.

		Predicted class			
		Autism	Normal	Parkinson's disease	Schizophrenia
Actual Class	Autism	6	0	4	0
	Normal	0	10	0	0
	Parkinson's Disease	8	0	2	0
	Schizophrenia	0	0	0	10

Table 7.3 Evaluation variables.

	Autism	Normal	Parkinson's disease	Schizophrenia
Accuracy	0.7	1	0.7	1
Sensitivity	0.6	1	0.2	1
Specificity	0.733	1	0.866	1
Precision	0.43	1	0.33	1
F1 score	0.57	1	0.52	1

7.4   Conclusion

The focus of this paper is to develop an efficient system to diagnose neural conditions using a suitable Deep Learning algorithm. The data acquired from training datasets recorded from different subjects and the results from testing data were used to calculate the output efficiency. The estimated accuracy rate was found to be 85% [28].

It is also observed that Autism and Parkinson's Disease classification were less precise when compared to the other classes. This indicates that with further work on the features extracted from Autism and Parkinson's Disease is required to improve the efficiency of the classifier. The algorithm can be implemented to more accurately perform Deep Learning and show better results. The increase in automation technology can reduce the workload of healthcare workers [29].

References

1. Abhishek Kumar & Jyotir Moy Chatterjee & Pramod Singh Rathore, 2020. "Smartphone Confrontational Applications and Security Issues," International Journal of Risk and Contingency Management (IJRCM), IGI Global, vol. 9(2), pages 1-18, April.
2. B.J. Bipin Nair, N. Shobha Rani, S. Saikrishna, C. Adith (2019), "Experiment to Classify Autism through Brain MRI Analysis", International Journal of Recent Technology and Engineering (IJRTE).
3. Bhargava, N., Bhargava, R., Rathore, P. S., & Kumar, A. (2020). Texture Recognition Using Gabor Filter for Extracting Feature Vectors With the Regression Mining Algorithm. International Journal of Risk and Contingency Management (IJRCM), 9(3), 31-44. doi:10.4018/IJRCM.2020070103.
4. C. Salvatore, A. Cerasa, I. Castiglioni, F. Gallivanone, A. Augimeri, M. Lopez, G. Arabia, M. Morelli, M.C. Gilardi, A. Quattrone (2013) "Deep learning on brain MRI data for differential diagnosis of Parkinson's disease and Progressive Supranuclear Palsy", Journal of Neuroscience methods.
5. Daljit Singh, Kamaljeet Kaur (2011), "Classification of Abnormalities in Brain MRI Images using GLCM, PCA, SVM", International Journal of Engineering and Advanced Technology.
6. David J. Brooks (2012), "Parkinson's disease: Diagnosis", Parkinsonism and Related Disorders Volume 18, Supplement 1, Pages S31-S33.
7. Donald J. Norris (2020), "Deep learning with the Raspberry Pi", Springer Science and Business Media LLC.

8. Franscisco Pereira, Tom Mitchell, Mathew Botvinick (2009), "Deep learning classifiers and fMRI: A tutorial Overview", NeuroImage Volume 45, Issue 1, Supplement 1, Pages S199-S209.

9. Hugo G. Schnack, Mireille Nieuwenhuis, Neeltje E.M. van Haren, Lucija Abramovic, Thomas W. Scheewe, Rachel M. Brouwer, Hilleke E. Hulshoff Pol, René S. Kahn (2013), "Can structural MRI aid in clinical classification? A Deep learning study in two independent samples of patients with schizophrenia, bipolar disorder and healthy subjects", Elseveir Journals.

10. Khushbu, Isha Vats (2017), "Otsu Image Segmentation Algorithm: A Review", International Journal of Innovative Research in Computer and Communication Engineering, Volume 5 Issue 6.

11. Kumar Sahwal, ..., Kishore, ..., Singh Rathore, ..., & Moy Chatterjee, ... (2018). An Advance Approach of Looping Technique for Image Encryption Using in Commuted Concept of ECC. International Journal Of Recent Advances In Signal & Image Processing, 2(1).

12. Kumar, A., Chatterjee, J. M., & Díaz, V. G. (2020). A novel hybrid approach of svm combined with nlp and probabilistic neural network for email phishing. International Journal of Electrical and Computer Engineering, 10(1), 486.

13. Madina Hamiane, Fatema Saeed (2017), "SVM Classification of MRI Brain Images for Computer-Assisted Diagnosis", International Journal of Electrical and Computer Engineering (IJECE).

14. Martha E. Shenton, Chandlee C. Dickey, Melissa Frumin and Robert W. McCarley 2010),"A review of MRI findings in Schizophrenia", Schizophrenia Research.

15. Masaya Misaki, Youn Kim, Peter A. Bandettini, Nikolaus Kriegeskorte (2010), "Comparison of Multivariate Classifiers and response normalizations for pattern-information fMRI", NeuroImage, Volume 53, Issue 1, Pages 103-118.

16. Mohd Fauzi Bin Othman, Noramalina Bt Abdullah, Nurul Fazrena Bt Kamal (2011) "MRI Brain Classification Using Support Vector Machine", 2011 Fourth International Conference on Modeling, Simulation and Applied Optimization, IEEE.

17. N Varuna Shree, T N R Kumar (2018), "Identification and Classification of brain tumor MRI images with feature extraction using DWT and probabilistic neural network", Brain Informatics 5, 23-30.

18. N. Bhargava, A. Kumar Sharma, A. Kumar and P. S. Rathoe, "An adaptive method for edge preserving denoising," *2017 2nd International Conference on Communication and Electronics Systems (ICCES)*, Coimbatore, 2017, pp. 600-604, doi: 10.1109/CESYS.2017.8321149.

19. N. Bhargava, S. Dayma, A. Kumar and P. Singh, "An approach for classification using simple CART algorithm in WEKA," *2017 11th International*

Conference on Intelligent Systems and Control (ISCO), Coimbatore, 2017, pp. 12-216, doi: 10.1109/ISCO.2017.7855983.
20. N. Bhargava, S. Sharma, R. Purohit and P. S. Rathore, "Prediction of recurrence cancer using J48 algorithm," 2017 2nd International Conference on Communication and Electronics Systems (ICCES), Coimbatore, 2017, pp. 386-390, doi: 10.1109/CESYS.2017.8321306.
21. National Institute of Neurological Disorders and Stroke https://ninds.nih.gov/.
22. Naveen Kumar, Prakarti Triwedi, Pramod Singh Rathore, "An Adaptive Approach for image adaptive watermarking using Elliptical curve cryptography (ECC)", First International Conference on Information Technology and Knowledge Management pp. 89–92, ISSN 2300-5963 ACSIS, Vol. 14 DOI: 10.15439/2018KM19.
23. P.G. Spetsieris, Y. Ma, V. Dhawan, J.R. Moeller, D. Eidelberg (2006), "Highly automated computer-aided diagnosis of neurological disorders using functional brain imaging", Proc. SPIE 6144, Medical Imaging 2006: Image processing.
24. Paolo Brambilla, Antonio Hardan, Stefania Ucelli di Nemi, Jorge Perez, Jair C. Soares, Francesco Barale (2003), "Brain anatomy and development in autism: review of structural MRI studies", Brain Research Bulletin 61 (2003) 557–569.
25. Rathore, P.S., Chatterjee, J.M., Kumar, A. et al. Energy-efficient cluster head selection.
26. Raymond Salvador, Joaquim Radua, Erick J Canales-Rodríguez, Aleix Solanes, Salvador Sarró, José M Goikolea, Alicia Valiente, Gemma C Monté, María Del Carmen Natividad, Amalia Guerrero-Pedraza, Noemí Moro, Paloma Fernández-Corcuera, Benedikt L Amann, Teresa Maristany, Eduard Vieta, Peter J McKenna, Edith Pomarol-Clotet (2017) "Evaluation of Deep learning Algorithms and Structural Features for Optimal MRI-based Diagnostic Prediction in Psychosis", Plos One Journal.
27. Rong Chen, Yun Jiao, Edward H Herskovits (2011), "Structural MRI in Autism Spectrum Disorder", Nature Articles, Pediatric Research Journal.
28. S. Bonavita, F. Di sale, G. Tedeschi (1999), "Proton MRS in neurological disorders", European Journal of Radiology.
29. Singh Rathore, P., Kumar, A., & Gracia-Diaz, V. (2020). A Holistic Methodology for Improved RFID Network Lifetime by Advanced Cluster Head Selection using Dragonfly Algorithm. International Journal Of Interactive Multimedia And Artificial Intelligence, 6 (Regular Issue), 8. http://doi.org/10.9781/ijimai.2020.05.003.

8

Convolutional Networks

Simran Kaur* and Rashmi Agrawal

Manav Rachna International Institute of Research and Studies, Faridabad, India

Abstract

Convolutional networks are known as convolutional neural networks when they have a grid like structure. Examples of CNNs include time series data which is seen as a 1-D grid taking samples at regular time intervals and image datasets which can be thought of as a 2-D grid of pixels. CNN is a deep neural network originally designed for image analysis. Recently, it was discovered that CNN also has an excellent capacity in consequent data analysis, such as natural language processing. CNN always contains two basic operations, namely convolution and pooling. A convolution operation using multiple filters is able to extract features (feature map) from the data set, through which their corresponding spatial information can be preserved. The pooling operation, also called sub-sampling, is used to reduce the dimensionality of feature maps from the convolution operation. Max pooling and average pooling are the most common pooling operations used in CNN. Due to the complicity of CNN, rely is the common choice for the activation function to transfer gradient in training by back propagation.

In this chapter, we define what a convolution is, its need in neural networks, and the application of pooling on different datasets. This chapter also addresses how to use a CNN and the kind of operations it applies on a dataset. In this chapter, CNN are applied on image assets for feature extraction and dimensionality reduction.

Keywords: CNN, rely, convolution, pooling, pixels

Corresponding author: Simrankaur1611@yahoo.com

Pramod Singh Rathore, Vishal Dutt, Rashmi Agrawal, Satya Murthy Sasubilli, and Srinivasa Rao Swarna (eds.) *Deep Learning Approaches to Cloud Security*, (109–122) © 2022 Scrivener Publishing LLC

8.1   Introduction

With the incredible achievement of profound learning in picture grouping, scientists began to investigate how to improve object discovery execution with profound learning. In recent years, object recognition has become dependent on profound learning and, likewise, has accomplished incredible advancements.

8.2   Convolution Operation

The convolution operation takes input as a tensor with shape and gives out an output with features, as well as with shape [1]. These feature maps are used to classify data. A convolution network convolves the input and passes it to the next layer. The following operations are performed on data:

A. **Pooling:** CNN may include local or global layers to carry on the underlying computation. It computes the maximum or average of a cluster of neurons.

B. **Fully Connected:** It is an MLP which connects one layer in the neuron to other layers in other neurons.

C. **Receptive Field:** The input area of a neuron is called the receptive field. The subarea of the original input image in the receptive field is increasingly growing as it gets deeper in the network architecture. This is due to applying a convolution which takes into account the value of a specific pixel, but also some surrounding pixels, over and over again.

D. **Weights:** The vector of weights and bias is called a filter and represents particular features of the input (e.g., a particular shape). A distinguishing feature of CNNs is that many neurons can share the same filter. This reduces memory trace because a single bias and a single vector of weights are used across all receptive fields sharing that filter, as opposed to each receptive field having its own bias and vector weighting.

8.3   CNN

To evade this issue, motivated by the association of the creature visual cortex, convolutional neural organizations (CNNs) were proposed to utilize the weight sharing system for abusing comparable structures occurring in various areas in a picture. Sharing the convolutional loads locally for a whole picture radically decreases the measure of boundaries that should be learned and renders the organization comparable concerning interpretations of the info (i.e., the quantity of loads no longer relies upon the size of the information pictured) [2]. The fundamental design of CNNs appearing in Figure 8.1 contains a few unique layers with different capacities.

The convolutional layers are the center structure squares of a CNN. At each convolutional layer, the information is convolved with a bunch of K learnable parts, W = {W1, W2, WK}, added by predispositions, b = {b1, b2,...,bK}. These portions are otherwise called open fields [3]. At that point, another component map, Xk, is produced by contributing the convolution results to a component insightful nonlinear capacity, σ(·). Given the yield vector of 1st layer, the kth include map is determined by

$$XL+1 = \sigma (WL * Xl + bl) \qquad (8.1)$$

The capacity, σ (·), is otherwise called actuation work and can be numerous capacities, for example, the sigmoid capacity σ(x) = (1 + e−x) −1, exaggerated digression σ(x) = tanh(x), or redressed straight units σ(x) = max (0, x). In spite of the fact that these bits in each convolutional layer have small responsive fields, they can be stretched out through the full profundity of the information volume. Stacking the component maps for all channels along the profundity measurement shapes the full yield volume of the convolutional layer [4].

Normally, the pooling layer follows the convolutional layer and plays out nonlinear down-examining. There are a few nonlinear capacities to execute down-examining activities for pooling layers, among which max pooling is the most well-known activity. It segments the component map into a bunch of non-overlapping square shapes and, for each such sub region, yields the greatest qualities. In notwithstanding max pooling, the pooling layer can likewise perform other nonlinear tasks, for example,

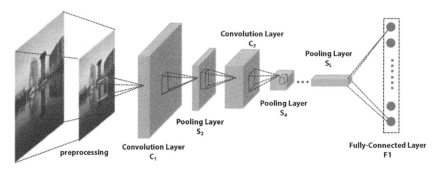

Figure 8.1 CNN.

normal pooling or L2-standard pooling. The pooling layer serves to logically lessen the spatial size of halfway portrayals, the quantity of boundaries, and the measure of calculation in CNN engineering and, subsequently, to control overfitting [5]. In addition, the pooling activity can give a type of interpretation invariance.

8.4   Practical Applications

Profound learning strategies have been broadly applied in numerous territories, such as PC vision and normal language handling, and are turning into the cutting edge techniques in these areas. As per the kinds of information, the application areas of profound learning will be sorted by sound information, picture information, and text information. Around the particular assignments, different profound learning approaches have been introduced to enormously grow the essential sorts of profound learning. All the more explicitly, these utilizations of profound learning allude to numerous intriguing issues, for example, discourse acknowledgment, face acknowledgment, object identification, scene order, and machine interpretation [6].

8.4.1   Audio Data

Sound information arrives in an assortment of structures, where discourse is the most widely recognized. In previous years, the discourse acknowledgment framework generally embraced the Gaussian Blend Model (GMM) to portray the likelihood models of each displaying unit. This model has, for quite some time, been involved in restraining infrastructure positions in discourse acknowledgment applications because of its straightforward assessment for mass information, just as the developed strategies for separation preparing. In any case, the GMM is basically a shallow organization and can't completely portray the qualities of the dissemination of state space [7]. Also, the component dimensionality of GMM demonstrating is commonly handfuls and makes it hard to depict the relationship between highlights. Event-partner, the GMM display, is basically a probability likelihood model. Despite the fact that segment preparing can impersonate the qualifications between certain classes, the capacity is, as of yet, restricted.

8.4.2   Image Data

In 2012, CNNs were right off the bat used in the Image Net Challenge for Picture Order and got best execution. In ensuing years, numerous CNN-based profound techniques have been introduced in the test and accomplished best in class results. Additionally, the methodology of numerous sliding windows is joined with CNNs to confine objects in the test and more profound structures with more customary procedures have been applied to this assignment. From 2014, all groups in the test have embraced profound learning strategies to rival each other. Contrasting the customary methodologies, the profound learning strategies involve an extraordinary favorable position in the undertakings of picture order and article location [8].

8.4.3   Text Data

The primary profound learning system was applied to traditional NLP issues by Colbert *et al.* in 2008. This system utilized 1D convolutional networks for all subtasks and accomplished cutting edge exhibitions. At that point, Cho *et al.* introduced an RNN-based structure with fixed-size vector portrayals for machine interpretation. In this profound model, two RNNs are incorporated: one RNN is utilized to encode a bunch of source language image groupings into a bunch of fixed-size vectors and the other RNN translates the vector into a bunch of representative successions of the objective language [9]. To beat the weaknesses of fixed-size vector portrayals, a profound model dependent on bidirectional RNNs was developed.

8.5   Challenges of Profound Models

Contrasted with other learning strategies, for example SVM, profound learning procedures are hard to decipher hypothetically. As the profound models are normally comprised of a few layers and measures of units, it is difficult to decipher nearby structures exclusively. Now and then, profound learning strategies are viewed as secret elements. Since profound models have a huge number of boundaries, the combination of preparing techniques cannot be examined without any problem. Particularly, a significant limitation is that the boundary introduction of profound adapting needs more exact direction in principle [10].

Contrasted with shallow models, profound models have all the more impressive capacity for portrayals with nonlinear capacity. Normally, for a self-assertive nonlinear capacity, both shallow organizations and profound organizations can be found for acceptable representations as indicated by the general estimation hypothesis of neural organizations. In any case, for a few capacities, profound organizations just need far less boundaries. Thus, we have to comprehend the example unpredictability that is the number of tests we have to gain proficiency with a decent profound model [11]. It is exceptionally hard to examine this hypothetically in view of the non-curved capacity.

On the opposite side, it is currently hard to plan a profound model for explicit errands. It is, by all accounts, that the normal preparing unit, similar to the convolution, exits in profound models for sound, picture, and text information. It is fascinating that there can be a bound together system built for any sort of information. Furthermore, in explicit assignments, it is currently a test to fuse space skills in these profound models [12].

Moreover, it is currently difficult to utilize the profound model to speak to organized semantic data. From a transformative perspective, language capacity is a long way behind visual and audible turn of events. So, from this point of view, it is hard to handle conceptual issues of language and effectively tackling this issue can help with the acknowledgment of fake general knowledge.

8.6   Deep Learning In Object Detection

With the incredible achievement of profound learning in picture grouping, scientists began to investigate how to improve object discovery execution with profound learning. In recent years, object recognition has become dependent on profound learning and, likewise, has accomplished incredible advancements. The mAP of item location on PASCAL VOC2007 significantly increments from 58% (in view of RCNN with Alex Net) to 86% (in light of Faster RCNN with ResNet). Right now, the best in class techniques for profound item location depend on profound convolutional neural organizations (CNN).

Here, some normal CNN designs for article recognition will be presented. Pedestrian detection, as a special case of object detection, will be explicitly examined.

8.7   CNN Architectures

Two-stage techniques treat object discovery as a multistage cycle. Given an information picture, a few recommendations of potential items are initially removed. From that point forward, these recommendations are additionally characterized into particular article classifications by the prepared classifier. The advantages of these techniques can be summed up as follows: (1) it lessens an enormous number of recommendations which are placed into the accompanying classifier, subsequently, it can quicken recognition speed; (2) the progression of proposition age can be viewed as a bootstrap method. In view of the recommendations of potential articles, the classifier can zero in on the arrangement task with little impact on the foundation (or simple negatives) in the preparation stage. Consequently, it can improve discovery exactness. Among these two-stage strategies, the arrangement of RCNN, including RCNN [12], SPPnet, Fast RCNN, and Faster RCNN, are extremely agent [13].

With the incredible achievement of profound convolutional neural organizations (CNN) on picture arrangement, Girshick *et al.* at first endeavored to apply profound CNN to protest discovery and proposed RCNN. In contrast to the conventional exceptionally tuned DPM, RCNN improves mean normal accuracy (mAP) by 21% on PASCAL VOC2010. Figure 8.2 shows the design of RCNN. It tends to be separated into three stages: (1) it initially removes the applicant object proposition, where the item recommendations are class autonomous and can be separated by the objectless techniques, for example, particular hunt, Edge Box, and BING; (2) for each article proposition of discretionary scale, the picture information is then distorted into a fixed size (e.g., 227×227) and put into the profound CNN organization (e.g., Alex Net) to figure a 4096-d highlight vector; (3) finally, in view of the component vector removed by CNN organization, the SVM classifiers foresee the particular classification of every proposition [14].

In the preparation stage, the item proposition ought to be initially produced by q specific request for preparing CNN organization and SVM classifiers. The CNN organization (e.g., Alex Net) is right off the bat pre-prepared on Image Net and afterward adjusted on the explicit article location dataset (e.g., PASCAL VOC). Since the quantity of the article class on Image Net and PASCAL VOC is extraordinary, the yields of the last completely associated layer in CNN organization ought to be changed from 1000 to 21 when tweaking on PASCAL VOC. The quantity of 21 speaks to 20 item classes of PASCAL VOC and the foundation. At the point when adjusting the CNN organization, the proposition is marked as the positive for the coordinated class in the event that it has the most extreme IoU cover with a ground-truth bouncing box and the cover is at any rate 0.5. Something else, the proposition is named as the foundation class. In view of the CNN highlights removed from the prepared CNN organization, straight SVM classifiers for various classes are additionally prepared individually [15]. When preparing the SVM classifier for each class, just the ground-truth jumping box is marked as positive. Something else, the proposition is named as a negative on the off chance that it has an IOU cover beneath 0.3 with all the ground-truth bouncing boxes. Since the separated CNN highlights are too enormous to even consider loading in memory, the bootstrap strategy is utilized to mine the hard negatives in preparing SVM classifiers.

To improve area exactness of the proposition, the straight relapse model is additionally prepared to anticipate a more precise bouncing box dependent on the pool5 highlights of the prepared CNN organization (i.e., Alex Net). Expecting that the first jumping box of the proposition (i.e., P) is spoken to by PX, Py, Pw, and Ph, where Px and Py are the directions of the focal point of the proposition P and Pw and Ph are the width and the stature of the proposition P, the bouncing box of the relating ground-truth (i.e., G) is spoken to by Gx, Gy, Gw, and Gh). At that point, the relapse focus for guaranteed (P, G) can be composed as:

$$TX = (Gx - Px)/Pw, \tag{8.2}$$

$$Ty = (Gy - Py)/Ph, \tag{8.3}$$

$$Tw = \log (Gw/Pw), \tag{8.4}$$

$$Th = \log (Gh/Ph). \tag{8.5}$$

To foresee the relapse target (i.e., $(t\hat{}, t\hat{}, t\hat{}, t\hat{})$) for another proposition (i.e., P), xywh, the pool5 highlights of the proposition spoke to as φ (P) are utilized. Along these lines, $t\hat{}$ (P) = 5$*$ w$*$T φ5 (P), where w$*$ is a learnable boundary and $*$ implies one of x, y, w, and h. Given the preparation test sets {(Pi, GI)}, where I = 1, 2, N. w$*$ can be improved by the regularized least squares objective as follows:

$$w = \text{argmin} (t\hat{} - w\hat{}T\varphi (P)) 2 + \lambda ||w\hat{}||2, \tag{8.6}$$

where λ is a regularization factor which is normally set as 1000. When preparing the relapse model for each class, the suggestion that has an IOU cover over 0.6 with a ground-truth jumping box is utilized. Something else, the proposition is disregarded [16]. In view of the scholarly w$*$ (i.e., wx, wy, ww, WH), the anticipated jumping box of proposition (P) can be determined as follows:

$$PX = PwwxT\varphi5 (P) + PX, \tag{8.7}$$

$$Py = PhwyT\varphi5 (P) + Py, \tag{8.8}$$

$$Pw = Pwexp (wwT\varphi5 (P)), \tag{8.9}$$

$$Ph = Ph \exp (whT \; \varphi5 \; (P)). \qquad (8.10)$$

The new anticipated recommendations will have a more exact area precision.

In spite of the fact that RCNN significantly improves item discovery execution, the article proposition ought to be twisted into a fixed size and afterward put into the CNN organization separately. Since the calculation of CNN highlights of various recommendations are not shared, RCNN is very tedious. To eliminate the fixed-size imperative and quicken location speed, Figure 8.2 analyzes RCNN and SPPnet. Rather than editing or twisting the picture information of the apparent multitude of propositions prior to processing the CNN highlights, SPPnet right off the bat registers all the convolutional highlights of the entire picture and afterward utilizes spatial pyramid pooling to extricate the fixed-size highlights of every proposition [17]. Figure 8.3 gives the illustration of the spatial pyramid pooling layer (SPP).

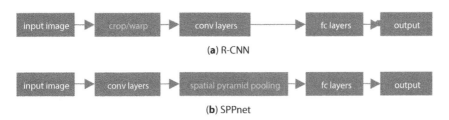

Figure 8.2 Comparison of RCNN and SPPnet.

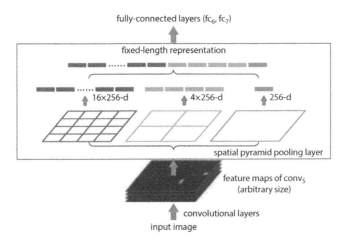

Figure 8.3 Pooling.

8.8   Challenges of Item Location

In spite of the fact that object identification has accomplished incredible advancements in the past few years, it actually has numerous difficulties when pushing the advancement of item discovery. In the accompanying section, three normal and average difficulties of article discovery will be examined and a few arrangements are additionally presented [18].

8.8.1   Scale Variation Problem

As the good ways from objects to camera can be different, objects of different scales typically show up on the picture. Along these lines, scale variety is an inescapable issue for object location. The answers for scale variety can be isolated into two principle classes: (1) picture pyramid-based strategies and (2) include pyramid-based techniques. By and large, picture pyramid-based strategies initially resize the first picture into various scales and afterward utilize similar indicators to distinguish the rescaled pictures, separately. Highlighted pyramid-based techniques initially produce numerous element guides of various goals dependent on the info picture and afterward utilize diverse element guides to identify objects of various scales [19].

From the outset, profound item location embraces the picture pyramid to distinguish objects of different scales. RCNN, SPPnet, Fast RCNN, and Faster RCNN all receive picture pyramids for object recognition. In the preparation stage, the CNN finder is prepared dependent on the pictures of a given scale. In the test stage, picture pyramids are utilized for multi-scale object identification. From one viewpoint, it, for the most part, causes irregularity between the preparation and test induction. Then again, each picture of picture pyramids is placed into the CNN organization, separately. Along these lines, it is likewise very tedious.

Truth be told, the element guides of various goals in CNN can be viewed as a component pyramid. On the off chance that the element guides of various goals are utilized to recognize objects of various scales, it can abstain from resizing the information picture, quickening recognition speed. Consequently, highlight pyramid-based techniques became famous. Analysts have undergone numerous endeavors pertaining to include pyramid-based strategies. Scale aware Fast RCNN (SAF RCNN) for person on foot location. The base organization is parted into two sub-networks for enormous scope walker identification and little scope passerby discovery, separately. Given an identification window, the last recognition score is the weight amount of two sub-organizations. In the event that the discovery window is generally huge, the huge scope network has a moderately huge weight. In the event that the recognition window is moderately small, the little scope network has a generally enormous weight. Yang *et al.* proposed scale-subordinate pooling (SDP) for multi-scale object discovery to deal with the scale variety issue. It depends on Fast RCNN architecture. The proposition is separated by the specific pursuit strategy. As per the statures of proposition, SDP pools the highlights of recommendations from various convolutional layers as per the tallness of proposition. On the off chance that the stature of item proposition has a place with [0, 64], SDP pools the component maps from the third convolutional blocks. On the off chance that the tallness of item proposition has a place with [18], SDP pools the component maps from the fourth convolutional blocks. In the event that the tallness of article proposition has a place with [128, +INF], SDP pools the component maps from the fifth convolutional blocks. Because the element guides of the ROI pooling layer are pooled from various convolutional layers, three distinctive subnets for grouping and finding proposition are prepared separately.

For the most part, Faster RCNN needs to remove recommendations by sliding RPN on a fixed convolutional layer (e.g., conv5_3 of VGG16). Since the separate field of a convolutional layer is moderately fixed, it cannot coordinate the spans of all articles very well. The particular field of the previous convolutional layer is generally small, which coordinates the little scope protests better, while the separate field of the last convolutional layer is moderately huge, which coordinates huge scope protests better. To tackle this issue, multi-scale profound convolutional neural organization is used (MS-CNN) to produce object recommendations of various scales. Figure 8.2 above shows the design of MS-CNN. It yields recommendations from various convolutional layers of various goals. The anchor in each convolutional layer is marked as the positive on the off chance that it has an IOU cover over 0.5 with a ground-truth jumping box. The anchor in each convolutional layer is marked as the negative in the event that it has an IOU cover beneath 0.2 with all the ground-truth bouncing boxes.

8.8.2   Occlusion Problem

Article impediment is exceptionally normal. For instance, it was discovered that most walkers (about 70%) in road scenes are impeded in, at any rate, one edge. Accordingly, distinguishing impeded articles is extremely vital and significant for PC vision application. In the previous decade, scientists have undergone numerous endeavors to tackle impediment issues.

It is found that if a few pieces of passerby are impeded, the square highlights of relating locale consistently react to the square scores of a straight classifier. In view of this wonder, they proposed to utilize the score of each square to decide if the relating locale is impeded. In light of the scores of each square, the impediment probability pictures are portioned by the mean move approach. On the off chance that impediment happens, the part identifier is applied on the unconcluded areas to yield the last location result.

To expand discovery execution on the impeded people on foot, it was proposed to gain proficiency with a bunch of impediment explicit person on foot finders. Every common locator serves for the impediment of a specific sort. Impediment can be separated into three unique sorts: impediments from base, impediments from right, and impediments from left. For each sort, the level of impediment goes from 0% to half. Eight remaining/right impediment identifiers and 16 base up impediment locators are prepared, separately. One nave strategy to get these classifiers is to prepare the classifiers of all the impediment levels individually. Be that as it may, it is very tedious. To diminish the preparation time, Franken-classifiers are proposed. They begin to prepare the full-body one-sided classifier and eliminate powerless classifiers to produce the principal impediment classifier. The extra frail classifiers of the first impediment classifier are additionally learned without predisposition. Like the main impediment classifier, the second impediment classifier is found out to be dependent on the principal impediment classifier. In view of Franken-classifiers, it just requires a one 10th calculation cost for preparing the arrangement of impediment explicit passerby identifiers.

8.8.3   Deformation Problem

Article misshaping can be brought about by non-lattice disfigurement, intra-class shape fluctuation, etc. For instance, individuals can bounce or hunch down. Accordingly, a decent item location strategy ought to be vigorous to the twisting of articles. Prior to CNN, analysts have undergone numerous endeavors to deal with twisting. For instance, DPM utilizes the combinations of multi-scale deformable part models (i.e., one low-goal root model and six high-goal part models) to deal with disfigurement. HSC further joins histograms of meager codes into deformable part models. To quicken the recognition speed of DPM, CDPM and FDPM are additionally proposed. Park *et al.* proposed to identify enormous scope walkers by deformable parts to demonstrate and recognize little scope people on foot by the unbending format. Regionlets present the district by a bunch of little sub-areas with various sizes and viewpoint proportions.

References

1. Abhishek Kumar & Jyotir Moy Chatterjee & Pramod Singh Rathore, 2020. "Smartphone Confrontational Applications and Security Issues," International Journal of Risk and Contingency Management (IJRCM), IGI Global, vol. 9(2), pages 1-18, April.

2. Bhargava, N., Bhargava, R., Rathore, P. S., & Kumar, A. (2020). Texture Recognition Using Gabor Filter for Extracting Feature Vectors With the Regression Mining Algorithm. International Journal of Risk and Contingency Management (IJRCM), 9(3), 31-44. doi:10.4018/ IJRCM. 2020070103.

3. Cai, Z. Fan, Q., Feris, R. S., and Vasconcelos, N.: A unified multi-scale deep convolutional neural network for fast object detection. in Proc. Eur. Conf. Comput. Vis. (2016).

4. Dollár, P., Wojek, C., Schiele, B., and Perona, P.: Pedestrian detection: An evaluation of the state of the art. IEEE Trans. Pattern Analysis and Machine Intelligence 34(4), 743–761 (2012).

5. Felzenszwalb, P. F., Girshick, R., and McAllester, D.: Cascade object detection with deformable part models. in Proc. IEEE Conf. Computer Vision and Pattern Recognition (2010).

6. Felzenszwalb, P., Girshick, R., McAllester, D., and Ramanan, D.: Object detection with discriminatively trained part based models. IEEE Trans. Pattern Analysis and Machine Intelligence 32(9), 1627–1645 (2012).

7. Girshick, R., Donahue, J., Darrell, T., and Malik, J.: Rich feature hierarchies for accurate object detection and semantic segmentation. in Proc. IEEE Conf. Computer Vision and Pattern Recognition (2014).

8. Kumar Sahwal, ..., Kishore, ..., Singh Rathore, ..., & Moy Chatetrjee, ... (2018). An Advance Approach of Looping Technique for Image Encryption Using in Commuted Concept of ECC. International Journal Of Recent Advances In Signal & Image Processing, 2(1).

9. Kumar, A., Chatterjee, J. M., & Díaz, V. G. (2020). A novel hybrid approach of svm combined with nlp and probabilistic neural network for email phishing. International Journal of Electrical and Computer Engineering, 10(1), 486.

10. Li, J., Liang, X., Shen, S., Xu, T., and Yan, S.: Scale-aware Fast R-CNN for pedestrian detection. CoRR abs/1510.08160 2015.

11. Mathias, M., Benenson, R., Timofte, R., and Van Gool, L.: Handling occlusions with franken- classifiers. in Proc. Int. Conf. Comput. Vis. (2013).

12. N. Bhargava, S. Dayma, A. Kumar and P. Singh, "An approach for classification using simple CART algorithm in WEKA," 2017 11th International Conference on Intelligent Systems and Control (ISCO), Coimbatore, 2017, pp. 212-216, doi: 10.1109/ISCO.2017.7855983.

13. Naveen Kumar, Prakarti Triwedi, Pramod Singh Rathore, "An Adaptive Approach for image adaptive watermarking using Elliptical curve cryptography (ECC)", First International Conference on Information Technology and

Knowledge Management pp. 89–92, ISSN 2300-5963 ACSIS, Vol. 14 DOI: 10.15439/2018KM19.

14. Park, D., Ramanan. D., and Fowlkes, C.: Multiresolution models for object detection. in Proc. Eur. Conf. Comput. Vis. (2010).

15. Rathore, P.S., Chatterjee, J.M., Kumar, A. *et al.* Energy-efficient cluster head selection through relay approach for WSN. J Supercomput (2021). https://doi.org/10.1007/s11227-020-03593-4.

16. Ren, S., He, K., Girshick, R., and Sun, J.: Faster R-CNN: Towards real-time object detection with region proposal networks. in Proc. Advances in Neural Information Processing Systems (2015).

17. Ren, X., and Ramanan, D.: Histograms of sparse codes for object detection. in Proc. IEEE Conf. Computer Vision and Pattern Recognition (2016).

18. Singh Rathore, P., Kumar, A., & Gracia-Diaz, V. (2020). A Holistic Methodology for Improved RFID Network Lifetime by Advanced Cluster Head Selection using Dragonfly Algorithm. International Journal Of Interactive Multimedia And Artificial Intelligence, 6 (Regular Issue), 8. http://doi.org/10.9781/ijimai.2020.05.003.

19. Wang, X., Han, T. X., and Yan, S.: An HOG-LBP human detector with partial occlusion handling. in Proc. IEEE Conf. Computer Vision and Pattern Recognition (2008).

9

Categorization of Cloud Computing & Deep Learning

Disha Shrmali

MDS University, Ajmer, India

Abstract

Cloud Computing is a technology that utilizes the web and remote servers to extract or store data and applications. Cloud Computing provides its services and resources in the form of pools of service models, namely software as a service (SaaS), platform as a service (PaaS), and infrastructure as a service (IaaS), whereas Deep Learning standards confront issues on which shallow models like SVM are highly affected by the menace of dimensionality. As a module of a two-phase learning plan counting numerous layers of non-straight management a lot of noticeably robust highpoints are logically unglued from the evidence. The existing instructional workout awarding ESANN in Deep Learning has unusual consultation delicacies compared to cutting-edge models and results in the present understanding of this learning method, which is an orientation for some difficult classification activities. So, in this chapter we will learn about how Cloud Computing and Deep Learning have taken over the world with their new and improved technologies and will learn about their applications, advantages, disadvantages, and correlations regarding different applications.

Keywords: Cloud computing, deep learning, CNN, ANN, RNN, SVM, etc.

9.1 Introduction to Cloud Computing

9.1.1 Cloud Computing

Cloud computing is the on-request accessibility of computer system resources, overwhelmingly information storage (cloud storage) and calculating control,

---

Email: dishashrimali799@gmail.com

Pramod Singh Rathore, Vishal Dutt, Rashmi Agrawal, Satya Murthy Sasubilli, and Srinivasa Rao Swarna (eds.) *Deep Learning Approaches to Cloud Security*, (123–144) © 2022 Scrivener Publishing LLC

deprived of straightway energetic administration by the client. This implies cloud computing is the conveyance of computational amenities counting servers, records schmoosing, software, storage, and intellect on the Internet to give quicker change, versatile assets, and economies of scale.

The articulation is frequently used to signify centres storing data available to numerous clients over the Web. High performance clouds generally have areas apportioned over various areas from focal workers. In the event that the relationship to the client is satisfactorily close, it may be designated as an edge worker [1].

Cloud computing hinges on the distribution of emanates to attain coherence and economies of high gage. People who favour public and hybrids get to circumvent or diminish front end costs. Cloud providers distinctively use a "pay-as-you-go" model, which can lead to fortuitous operating outlay if supervisors are not acquainted with cloud-pricing models.

The availability of high-proportion networks, low-cost computers, and storage devices, as well as the widespread endorsement of hardware virtualization, service-oriented architecture and autonomic, and utility computing has led to a great hike in cloud computing.

The elemental concept behind cloud computing is that the site of the service and many of the attributes, such as the hardware or operating system on which it is performing operations, are large bases tangential to the user. Keeping in mind that the metaphor of the cloud was taken from old telecom network schematics, in which the communal telephone network (and later the internet) was often denoted as a cloud to represent that didn't have major importance; it was just like a cloud of stuff. This is an oversimplification of course; for many customers location of their services and data remains a major issue [2].

9.1.2   Cloud Computing: History and Evolution

In the emerging age of IT, the Client-Server model was immensely welcomed along with the workstation and fatal apps. Storage of data in the Central Processing Unit at that moment was very exorbitant and the workstation linked all the models and presented their services to narrow passages approaching the client. As the mass storage capacity program was introduced, many of the companies providing such services gained a huge popularity for storing big lengths of series of data and information.

In the year 1990, the concept of internet had sufficient machines linked to it and the linking of those machines altogether made a gigantic pool of storage that hadn't been accomplished by a single organization or institution. Hence, a new concept of the "grid" was introduced. The term "grid" was considered to be a part of cloud computing as both the technologies were shaped from linking a web of computers to form a single chain connection. The grid obligates the use of application-oriented programs to split a single system of tasks into thousands of smaller machines. The only drawback grid computing faced was that if one major node flops to work, all the other nodes might also flop to execute a process. Therefore, grid computing was not able to prove itself very productive [3].

When we see the other face of the coin, cloud computing acquired and worked over the shadow of major notions of the grid; one of them being providing resources and services only when the demands arrived from the user's end [4].

The first milestone of cloud technology was accomplished when, in 1999, salesforce.com inscribed its name. It launched a method of transporting innovative apps through a website. They encompassed both dedicated and conventional IT firms to showcase themselves.

The second step was taken by Amazon Web services around mid-2002 when they attempted to publicize cloud services with the help of the Amazon Mechanical Turk [5]. After this, it launched Elastic Cloud Compute which acted as a web service to provide a pace to establish their businesses and could easily run on their desktops. The 1st cloud accessible to everyone was launched and named EC2-S3. This was almost the same time when Google launched Web 2.0, where Google, as well as many other companies, collaboratively started browser-based services on the same panel.

And then finally, came Microsoft's Azure where both Microsoft and Google delivered services in a profound way to make it more reliable and easier to acquire data, taking the Web to new heights.

9.1.3 Working of Cloud

The Cloud is a pool of space that is used to be filled with data that can be shared through the internet. A cloud application is fundamentally comprised of two layers. One is a host and the other is a hosting organisation. These hosting organisations are responsible for managing big data centres that deliver security, computational fuel, and storage capacity [6].

To understand how a cloud system works, it must be divided into two sections: the front end and the back end. They are attached to one another by a network, typically the Internet. The front end is the side of the computer user or client. The back end is 'the cloud' section of the system [7].

All the layers play their parts and work together as a combined architecture in a form of hierarchy that means that the output of one is taken as an input to the other. Its stats work with the front end, for example the

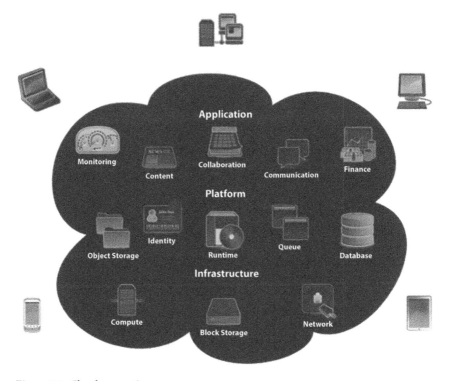

Figure 9.1 Cloud computing.

client's desktop panel, network, or a user interface connecting the user to the cloud, etc. The back end of the architecture works to expand computers, storage, servers, etc. in a diversified way shown in Figure 9.1.

Nowadays, everything from data processing to high bandwidth videogames relies on the cloud, where each type or application comprises of its own compatible server. Though many large companies and brands are moving towards the cloud, there is a profusion of other firms as well. Companies can sell their rights to big cloud firms so that they can store their data on the company's cloud, for example, when you download a song from any website, it automatically appears in your iTunes folder [8].

Generally, cloud computing follows three delivery models:

1. **Public:** The public model is the main model, which comes with gigantic contributions. Following ecosystems are presented to customers and are reachable by the public cyberspace. Customers do not need to be worried about any organisational proprietorship.

2. **Private:** The private cloud is different from a linguistic foundational information bureau. Both the employees and resources must acquire a datacentre and private cloud. The private model inherits pliability similar to a public cloud. However, in a private cloud the ecosystem is designed to have similar resource sharing and scalability as the public cloud, but with better security because only the owners can access this particular space.

3. **Hybrid:** This delivery model comprises of the features and specifications of both public and private cloud models. In this model, both clouds are inter-connected to each other through the Internet and alternatively provide resources according to the client's demand; that means that those responsibilities that private cloud fails to fulfil or get disabled, can be fixed with the help of a public cloud. This kind of technology proved to be prominent in situations like disaster recovery.

9.1.4 Characteristics of Cloud Computing

1. On-Demand Self-Services:
Cloud computing services do not require the assistance of any human being to handle, evaluate, observe, or administrate computing resources and can do these easily by themselves according to their requirement.

2. Broad Network Access:
Many users can access resources and services at the same time due to the availability of a broad network panel.

3. Rapid Elasticity:
The services that go through computational algorithms regularly require certain information technology-based resources that can quickly identify pros and cons as required. It is activated when a user requests a service and provides a result as soon as the demand is fulfilled.

4. Resource Pooling:
The IT resources presented are publicized transversely over several submissions and are occupied in a disassociated way. Multiple clients are provided service from the same physical resource.

5. Measured Service:
The pattern of consumption of resources is measured for every submission and occupant and it provides both the user and the resource provider with a detailed report of what has been used. This is done for various reasons, like monitoring, billing, and efficient use of resources.

9.1.5   Different Types of Cloud Computing Service Models

Models in cloud computing were formed to provide clients with pool configurable resources that they can use at their convenience. These can be servers, applications, network services, storage, configuration tools, etc. [9]. These services can be provided as fast as possible and can be restricted or publicized with minimal management efforts. Cloud Computing has evolved since the day it was first invented and through its journey it has distributed itself into three major service models:
   Infrastructure as a Service (IAAS)

- Platform as a Service (PAAS)
- Software as a Service (SAAS)

Let's learn more about these service models:

9.1.5.1   Infrastructure as A Service (IAAS)

When we work on a project, we require many resources to complete our tasks before a deadline, but the only resource to do any work is a virtualised computational resource and this prominent resource is provided by an IaaS model. An IaaS is basically a type of model where a third party in the client's space provides different types of amenities such as hardware, software, infrastructural components, etc. According to these requirements, IaaS providers broadcast the applications according to their requirements and resolve tasks, such as maintaining system backup or planning flexibility, one by one [10]. IaaS providers provide resources that are highly ascendable and can be manipulated according to the user's need. Because of these qualities, many users use these resources for experimental or temporary purposes. The aura of IaaS also comprises of automated admin tasks, visualisation of desktop, and services that are based on policies. The IaaS market is not very large yet and requires constant finances for its further development. In this task, Open Stack is providing its best services to accomplish every goal a cloud-based organisation might be looking forward to accomplishing.

IAAS Network
Load Balancing and Domain Name System are the two major network service providers presented to the world by the public cloud. Domain Name System works over computers, gaming devices, etc. by providing them with a unique hierarchical identification system. IP addresses are used by DNS to detect networks [11]. Load balancing, in contradiction, to DNS

creates a sing point of access for multiple users. It is done with the help of some load balancing algorithms, hence, we can conclude by saying that a load balancer is a networking set of devices that distributes network traffic within multiple servers using specified load balancing algorithms.

9.1.5.2   Platform as a Service (PAAS)

When we work in an IT organisation, we know it is basic to require application enragement or up gradation from time to time. The PaaS or Platform as a Service model provides hardware and software over the internet to clients from its own infrastructure. That means that PaaS grant their employees power to install in-house applications of hardware and software so that they could develop or execute new applications.

Your business is not undertaken by PaaS providers, but it is majorly dependent on its services like hosting of relevant applications or Java amplification. To do all this, the PaaS provider starts the procedure by logging in through the web. Then, he provides a charge sheet to the user according to the time period on a user-involved basis and then, according to his needs, he activates his services [12].

Important physiognomies of a PAAS model can be:

- Scalability
- Automated delivering of fundamental resources
- Safety
- Redundancy

9.1.5.3   Software as a Service (SAAS)

SAAS, as the name suggests, is a type of model that is being majorly used as a software distribution system in which the apps are accommodated by the service providing companies and make them accessible to their customers over the internet. With its increasing demands, SaaS has become the most protuberant delivery model, similar to the SOA and web services supported by fundamental technologies. Some newly launched technologies, which work from the same roots, have become in high demand as well. Some of the newly in demand service models or models like ASP, reflect the shadow of SaaS in their output shown in Figure 9.2 [13].

Some of the important points of using an SAAS model are stated below:

- Effortless supervision
- Involuntary updates and covering administration

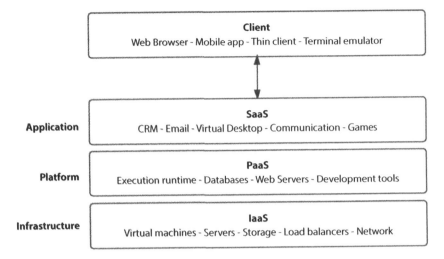

Figure 9.2 Service models.

- In terms of compatibility, all the clients have the same updated version of software
- Strong partnership because of the same motive
- World-wide approachability

9.1.6 Cloud Computing Advantages and Disadvantages

Cloud Computing is moulding our life and work nowadays. Whether we appreciate it or not, it has turned out to be an inherent fragment of our life. Highly marked corporations and industries of all kinds are now investing their money into Cloud Computing. But, as we all know, nothing is 100% perfect and Cloud Computing is not an anomaly. Though it is highly advantageous, it comprises approximate key jeopardies and anxieties that must not be neglected. Hence, let us deliberate the benefits and drawbacks of Cloud Computing in detail.

9.1.6.1 Advantages of Cloud Computing

Cloud Computing is an emerging technology that nearly every corporation is switching to for its foundation technologies. Let us look over the following advantages to know why cloud is favoured over other foundational technologies shown in Figure 9.3.

- **Cost Efficiency:** The major reason why companies are diverting their interest is because regardless of on-premise technology, Cloud Computing receipts significantly minimal cost. Because of this, companies need not store information on CDs anymore; the cloud proposes gigantic storage and interplanetary, redeemable money and resources.

- **High Speed:** Cloud Computing allows us to establish services swiftly with minimal exertion. This swift placement permits access to the possessions obligatory for the apparatus in a matter of seconds.
- **Excellent Accessibility:** In the heavy shadow of the cloud, we can access our data from anywhere regardless of location and portable machines make it easier to access technology.
- **Back-up and Restore Data:** Cloud Computing provides great backup support. Rather than other devices, once the data gets stored in the cloud, one can easily create a backup of it and stay free from the pressure of data loss.
- **Manageability:** Under the standard rules of SLA (Service Level Agreement), Cloud Computing makes sure that each and every client receive a seamless, on time guaranteed service delivery and takes care to manage and maintain

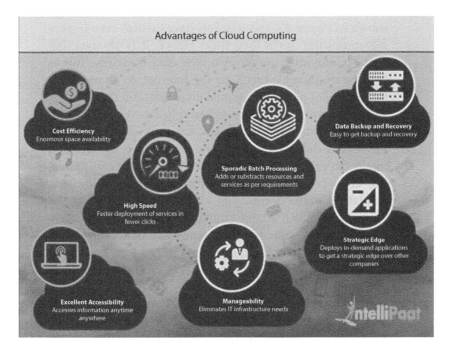

Figure 9.3 Advantages of cloud computing.

all the IT based apps and services, making it more reliable for clients.

- **Sporadic Batch Processing:** We can add or subtract any resources we like according to our need in a cloud environment. This means that if we do not work on fixed time periods, we will not have to pay for the services; we can easily deduct them for the time period and add them again when require.
- **Strategic Edge:** Cloud Computing provides a pack of all the latest, time oriented and pocket friendly applications to steal the spotlight from your contenders. They try to motivate their client organisations to beat everyone by taking over doing all the manual work and letting them focus on the competition by providing access to most gravitated apps.

9.1.6.2   Disadvantages of Cloud Computing

Every technology has both +ve and -ve facets that are highly essential and should be debated beforehand implementation. The points mentioned above culminate advantages of making use of cloud services and now, the discussion given below will outline the conceivable disadvantages of Cloud Computing.

- **Vulnerability to Attacks:** Cloud is a very new technology shown in Figure 9.4 and attackers have existed, even in very big companies, for a long time. Hence, it is obvious that Cloud Computing might also face data breaches. This is because the cloud is totally on the Internet, so an organisation must not put their essential documents on the cloud.
- **Network Connectivity Dependency:** In the corporate world, there are many days when work has to be done while the Internet is down for quite some time. Since the cloud is totally dependent on Internet, the absence of it can cause a loss in the company because they cannot access their essential data from the cloud.
- **Downtime:** When a large amount of people tries to access a single server, some technical issues like buffering, loss of power, lack of internet connectivity, speed loss, etc. might occur. This creates a downtime in services. Downtime is one of the major glitches for cloud users.
- **Vendor Lock-In:** When a situation occurs when one needs to switch from one cloud service to another, a company faces some major issues. This happens because the vendor platform might not provide you with the proper controls

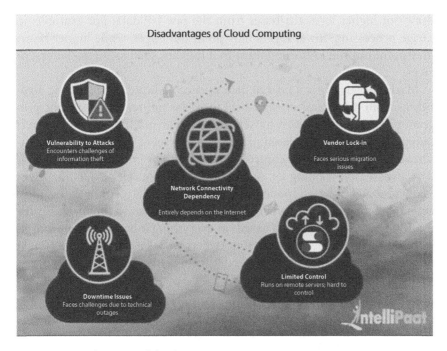

Figure 9.4 Disadvantages of cloud computing.

you require, it might create support and reliability issues, additional expenses for support tools might be requested, configuration complexities might occur, or, in a worst-case scenario, a company's essential data might get lost or be left unattended, making it vulnerable for attackers to breach it.

- **Limited Control:** Companies who generally work under wide shadows of the cloud face the issue of not getting 100% access to all the liabilities of the cloud. This is due to the fact that cloud services provide your own maximum of cloud servers which are situated in remote areas. Due to this, they cannot give certain controls to the user, which makes the client feel like they have limited control issue.

9.2   Introduction to Deep Learning

Deep Learning is a superior branch of the ML family that creates an environment of artificial neural networks in a computer with representative learning. In technological terms, we can define deep learning as a class of machine learning algorithms that uses multiple layers to progressively draw out higher level attributes from the raw I/P data. For example, in image processing, lower layers may point out edges, while higher layers may point out the notion relevant to a human, such as digits, letters, or faces.

Machine learning is a field of data science dedicated to acquiring algorithms from data based on statistical formulas not requiring specific programming. Machines learn by sending another machine a set of correct outputs, which are considered to be given as a set of inputs (i.e., what features are important and to control for), and then by providing the machine with an extensive training set. As time goes on, the machine tightens one of many probability-based models and engenders the algorithm that is most acceptable for data. Now, when new data is provided to the machine, the machine makes a prediction based on everything it has acquired from its learnings. This is supervised learning, where we tell a computer if it is right or not, but there is also unsupervised learning [14].

9.2.1   History and Revolution of Deep Learning

The first two words in the deep learning's book of history are "Warren McCulloch and Walter Pitts". These were the two men who gave life to the dream of making a computer based neural network inspired from the workings of an actual human brain. They tied mathematical equations with computer algorithms to create a thought process for the computer. After that, as history suggests, the lifecycle of Deep Learning reached great new heights.

Descriptions of some of these prominent achievements and achievers are mentioned below.

9.2.1.1   Development of Deep Learning Algorithms

In 1965, Grigoriev Litvinenko and Valentin Grigoriev Lapa took the first step towards the long journey in the development of Deep Learning. They used polynomial functions that were used to analyse and solve problems statically. After that, a disturbance occurred in the development of AI due to a lack of funding, yet the enthusiasm and curiosity of the team did not fall and they continued to do research in the field without getting any pay checks.

The name Kunihiko Fukushima was written with golden ink in the history of Deep Learning when he proposed a design for a neural network with different types of pooling and convolutional branches. He developed an ANN "Neocognitron" in 1979 by using hierarchical and multiple layered designs through which a computer was itself able to recognise, detect, or memorise visual patterns [15].

9.2.1.2   The FORTRAN Code for Back Propagation

The 1970's was when the evolution of the Back Propagation algorithm took place. This was the type of algorithm used to provide training to DL models. Seppo Linnainmaa introduced BPA with its FORTRAIN code and gained the attention of whole world. Though it grew it branches in the 1970's, it was not introduced to the neural network until 1985 when Rumelhart and Hinton Williams took a step forward. This step resulted into a great outbreak of many interesting, distributed representations [16]. In 1989, at Bell Labs, Yann LeCun expounded the first practical application of back propagation. In the application, he joined the two major forces, conventional neural networks and a Back Propagation algorithm, so that it could easily scan and read handwritten digits.

A big bang occurred between 1985 and 1990, when Vladimir Vapnik and Dana Cortes created a support system for the analysis and road mapping of data corresponding to each other. In 1997, Juergen Schmid Huber and Sepp Hoch Reiter introduced LSTM to implement RNN.

After that, in 1999, when computers grasped high speed GDPU processing, the speed of the computer was preemanated by 1000 times in a lifecycle of 10 years. This time period announced the beginning of neural networks versus support vector machines combat.

9.2.1.3   Deep Learning from the 2000s and Beyond

The beginning of the new century introduced Deep Learning to new heights of progress. In the early 2000s, officials came across the Vanishing Gradient Problem. In this problem, the upper layers of architecture were not able to recognise the features of lower levels due to failures caused in the detection of learning signals. This problem was restricted to only gradient-based learning methods. Due to this problem, the input data's range prominently increases while the range of output decreases, putting it into a situation of chaos. The second outbreak happened when the META group presented a paper on 3D data growth opportunities and obstacles. This research stated the hostility of big data and explained how speed and volume of data are directly proportional to data source and types. After this, in 2009, scientists and AI professionals, under the guidance of Fei-Fei Li, accumulated a data base independent in nature carrying around fourteen million images sorted by category. These images were introduced to the system as an input to provide recognition training to the neural network [17]. Then, until 2011, this segment was updated by training the network without the requirement of pretraining from layer to layer. In terms of speed and efficiency, Deep Learning has stated itself to very high standards.

9.2.1.4   The Cat Experiment

The cat experiment was an astonishing development in the field of Deep Learning, which was introduced to the world by "goggle brains" in 2012. It was a very basic yet innovative experiment which explored and demonstrated the issues we face while working with unsupervised learning. In this experiment, around 1000 computers were interlinked as a neural network from ten million random images. The images selected from YouTube

Figure 9.5 Experiments.

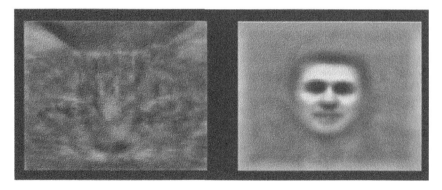

Figure 9.6 Cat experiment.

were randomly presented into sets. One set is of cats and the other one is of faces with no proper information provided for the image about its description or name. These two sets were then presented as an input to the software. The software trained itself by analysing those thousands of images to determine what a cat looks like versus what a man looks like. It also provides a picture of the nearest probability to be a cat or a face shown in Figure 9.5 and Figure 9.6.

9.2.2   Neural Networks

9.2.2.1   Artificial Neural Networks

Artificial Neural Networks or ANN are basically a type of neural system that processes stimulated data inspired by biological neural networks of the human brain. This system learns to accomplish a task by doing proper planning without the help of task-oriented programming. Let us consider an experiment where the system is trained to identify a turtle by inspecting thousands of images that contains turtles and another animal. In this situation, if we provide an image to the machine, then by scanning it can easily identify if the image is of a turtle or not. These applications are generally used in situations where users suffer difficulty to express themselves properly to computer algorithms. Over time, as the requirements of the user increased, ANN also updated itself by focusing attention on matching specific mental abilities, guiding divergence from biology such as back propagation, or passing information in the opposite direction and accommodating the network to reflect that information.

ANN is made of small groups of artificial neurons. Each neuron consists of a bridge in between one another that helps to cast signals from one neuron to another. The neuron at the receiver's end extracts the resultant of the signal and then casts it again to a neuron at lower levels. Neurons might consist of some value ranging in between 0-1 counted as a real number; it also carries a weight which determines the strength of the signal.

Neural networks are still being used to accomplish many complex tasks, namely speech recognition, introducing filtering in social networking sites, board and video games, and medical diagnosis. Since 2017, ANN consisted of bridges ranging from thousands to millions. Though these numbers are a lot less than the number of neurons present in the human brain, they can perform a lot of complex tasks a lot faster than a human brain could ever do [18].

9.2.2.2   Deep Neural Networks

A Deep Neural Network (or DNN), as the name suggests, is a deep study of neural networks where there are multiple layers interlinked to each other in between the input and output layers of the network. Many verified mathematical equations and manipulations were discovered by DNN so that it could obtain an intensified output by applying them over linear or nonlinear input data. When the network travels through every single layer, it evaluates the probability of each layer to determine the output. The perfect example of this is a DNN trained to identify a species of a spider; if we provide an image of a spider to the system to identify its species, to accomplish the following task, it will firstly scan the image and then it will try to calculate the probability of the spider in the image matches with the type of species present in the database. After getting the results, the user can set the limit of probability the machine must display.

A DNN is programmed to generate complex non-linear correlations; that means that its system creates models where the object is represented as primitive layer of configuration. The layer added as an extra sector takes the components and attributes from lower layers which are probably modelling with units that are executing on a low-level network. Many variables with basic approaches are included in deep architecture where each and every architecture is successful in a specific area. Reversing the performance of any kind of architecture is not yet concluded to be possible unless it has been evaluated a similar data set.

DNN network based data flow does not contain any loops when it flows between input and output layers. The steps indicating data flow in DNN are as follows: firstly, DNN generates a map of virtual neurons, then, it allots a random value to each relation coexisting between them [19]. These values and inputs are then multiplied, which returns the value between a range of 0-1. If a network fails to detect the pattern provided by the data, the algorithm readjusts the random value provided to the relation between vectors. This process of authentication of algorithms continues until an appropriate result is obtained.

9.2.3   Applications of Deep Learning

9.2.3.1   Automatic Speech Recognition

In the field of Deep Learning, the most prominent application ever made was for automatic speech recognition. It is the most persuasive and fortunately skilled method for speech recognition at a very large scale. To acknowledge multiple second time zone speech recognition, which is distributed into different time steps, we use LSTM based RNNs. Here, in LSTM, a single step is equivalent to ten ms and it is for a gate that does not support traditional methods of speech recognition for some specific tasks.

The recognition of the application started from a very small-scale point of view which was originated by TIMIT. Here, this dataset was used to carry around six hundred and thirty speakers, having knowledge of around eight regional language of America. Here, for every single speech creation, a person reads ten sentences. As it is very small in size, the device can configure itself as many times as it can. Moreover, TIMIT focuses on phone sequence identification rather than word sequence recognition to present a weak phone with a higher RAM language model. This results in the incrimination of strength in the traditional method of speech recognition and becomes a lot easier to be analysed. In today's world, all major speech recognition models are present at a commercial level, like Xbox, Apple Siri, Alexa, etc., are based on DL.

9.2.3.2   Electromyography (EMG) Recognition

Identification of a user's purpose to perform a certain task is a bit of a difficult analogy. There are many situations in a man's life where he needs to fully depend on a machine; in such a case, a machine needs to understand the actual purpose of the client to perform this command. This task of recognising the purpose of a user to acquire control of the device is conducted by Electromyography (EMG). Smart wheelchairs, prosthetic devices, and exoskeletons are certain examples that rely on the following technology.

In the beginning, era-dense neural networks were in business, but then, researchers, with the help of a spectrogram, mapped EMG signals and then used that info to provide it as an input to deep convolutional neural networks. Recently, core to core Deep Learning is being used to trace signals so that it can directly identify a user's purpose.

9.2.3.3   Image Recognition

MNIST is basically an examination conducted over images to evaluate and characterise their patterns. It is a concoction of around sixty thousand handwritten digits and around ten thousand examples of tests conducted on them. In the case of TIMIT, its small size becomes a prominent factor to allow the user to perform multiple tests over it. Deep Learning has proven itself to produce more accurate results in image recognition than a human itself. One such example is Facial Dysmorphology Novel Analysis.

9.2.3.4   Visual Art Processing

The range of advancement in the applications of Deep Learning has reached its heights by introducing the technique of visual image processing to the world. In this, DNN has played a major role by providing a gateway for neural style transfer for scanning of a style of provided art to apply it in a visually oriented manner to a whimsical picture or video technology, determining factual information regarding the artwork and originating striking imagery with reference to randomly chosen visual input fields.

9.2.3.5   Natural Language Processing

Language processing and implementation has been the greatest achievement in the field of neural networks. The study began in early 2000 where LSTM was used to improvise interpretation capacity of machine and language modelling.

There are some key proficiencies present in this operation that can lead us to accomplishing our task of language interpretation. In the Deep Learning formation, word embedding works as a representational layer that helps recreate an atomic word, representing it in the form of a structure relevant to other words in the dataset. The location of these datasets in a vector is known as point. Word embedding with the help of RNN or an input layer is used to parse phrases and large sentences. The major technology behind this is productive compositional vector grammar. At the top of word embedding, an auto encoder has been built to detect paraphrasing and gain similarity in a sentence. Grammar that comes in the goner of a compositional based vector is considered PCFG (Probabilistic Context Free Grammar), which is presented by RNN [20].

There have been many recent developments that have given us better ways to generalise sentence embedding from word embedding. Google Translate and Google Neural Machine Translation use the following techniques to make the machine independent for reading and interpret words by learning from millions of examples, carrying knowledge of hundreds languages, and encoding the semantics of sentences at the place of phrase-to-phrase memorisation and translating all.

9.2.3.6   Drug Discovery and Toxicology

High levels of competitor medications flop to win administrative endorsement. This kind of disappointment is brought about due to a lack of efficacy, unlonging connections, or uninitiated lethal impacts. Investigators examined the utilization of DL to figuring out how to foresee bimolecular focuses, off-targets, and poisonous impacts of ecological synthetic compounds in supplements, family items, and medications.

AtomNet is a Deep Learning framework for architectural judicious medication graph icing. AtomNet is utilized to foresee original up-and-comer biomolecules to disorder targets, for example, the Ebola haemorrhagic fever virus. Generative neural systems in 2019 were utilized so that they could deliver particles approved tentatively right into mice.

9.2.3.7   Customer Relationship Management

Profound fortification erudition is being utilize to surmise the estimation of conceivable direct advertising activities characterized regarding RFM factors. The evaluated esteem work appeared to have a characteristic understanding of client lifetime esteem.

9.2.3.8   Recommendation Systems

Proposal frameworks have utilized deep learning to figuring out how to separate important highlights to recommend better music, data, or journals according to certain requirements. For taking in client inclinations from multiple domains, multi-view deep learning is applied. This model utilises a multidimensional collaboration, content-based methodology, and upgrade suggestions in various errands.

9.2.3.9   Bioinformatics

To anticipate gene ontology observations and gene-function associations, we utilise an auto encoder ANN in bioinformatics.

9.2.3.10   Medical Image Analysis

Deep Learning has produced prominent results to showcase its high grip in medical based application in which lesion detection, segmentation of organs, characterisation of cancer cells, and enhancement of the image is conducted.

9.2.3.11   Mobile Advertising

The search for a suitable transportable audience to accomplish the purpose of versatile promoting still needs to be tested in this zone. An objective section can be made and utilized in advertisement serving by any promotion worker once the focuses of all the information are thoroughly analysed. Deep Learning is being utilized to decipher publicizing enormous multidimensional datasets. Numerous data focuses have been gathered during the request - serve - click web publicizing loop. The obtained data shapes the premise of AI to recover promotion determination [21].

9.2.3.12   Image Restoration

Issues such as denoising, higher resolution, inpainting, and film colorization are the kind of reverse order Deep Learning is applying over such issues to minimise its effectiveness. The following apps incorporate the understanding technique "Contraction Fields for Operative Image Refurbishment", which trains on a picture.

9.2.3.13   Financial Fraud Detection

Deep Learning acts as an effective platform to detect financial fraud and money laundering. "To spot and perceive connections and likenesses among information and figure out how to distinguish irregularities in the occurrence of specific events deep anti-money laundering recognition framework works wonders" [17]. This arrangement utilises both administered and non-administered learning techniques.

9.2.3.14   Military

Many countries defence departments have started applying Deep Learning approaches to create new ideas in the field of defence.

9.3   Conclusion

After going through this chapter, we can precisely conclude that Cloud Computing and Deep Learning both are very recent technological expansions that have the ability to have a great impact on the world. Cloud Computing has many benefits, such as delivery to users and businesses, providing great assistance to businesses, and decreasing functioning charges by having fewer expenditures on preservation and software promotions and emphasis more on the businesses itself. But, with the anonymous number of benefits, there are many challenges as well that Cloud Computing must overcome. Deep Learning is a part of the machine learning family that provides training to computers to do what originates automatically in human brains. In Deep Learning, a computer model attempts to achieve classification responsibilities on its own by scanning images, text, or sound. It is crucial to input a voice controller in client devices like hands-free speakers, TVs, tablets, and phones [22]. Deep Learning models can attain a state of art precision, occasionally surpassing human-level presentation. Models are skilled by utilising a great set of categorized data and ANN that encompass anonymous branches. Deep Learning is receiving great consideration latterly and for beneficial goals. Both of these technologies are superior and fundamentally useful in their own ways. They both have benefits and disadvantages and are providing benefits to society with their numerous applications.

References

1. A cloud-edge based data security architecture for sharing and analysing cyber threat information.
2. Abhishek Kumar & Jyotir Moy Chatterjee & Pramod Singh Rathore, 2020. "Smartphone Confrontational Applications and Security Issues," International Journal of Risk and Contingency Management (IJRCM), IGI Global, vol. 9(2), pages 1-18, April.
3. Bhargava, N., Bhargava, R., Rathore, P. S., & Kumar, A. (2020). Texture Recognition Using Gabor Filter for Extracting Feature Vectors With the Regression Mining Algorithm. International Journal of Risk and Contingency Management (IJRCM), 9(3), 31-44. doi:10.4018/IJRCM.2020070103
4. By Keith D. Foote :A Brief History of Deep Learning https://www.dataversity.net/brief-history-deep-learning/
5. Cloud Flare https://www.cloudflare.com/learning/security/what-is-a-fire wall/
6. Daniel S. Berman, Anna L. Buczak *, Jeffrey S. Chavis and Cherita L. Corbett A Survey of Deep Learning Methods for Cyber Security Mohamed Amine Ferrag, Leandros Maglaras, Sotiris Moschoyiannis, HelgeJanicke Deep learning for cyber security intrusion detection: Approaches, datasets, and comparative study
7. David W Chadwick ,Wenjun Fan, Gianpiero Costantino, Rogeriode Lemos, Francesco Di Cerbo, Ian Herwono, Mirko Manea, Paolo Mori, Ali Sajjad, Xiao-SiWang.
8. Deep learning architectures in emerging cloud computing architectures: Recent development, challenges and next research trend
9. Dhirendra KR Shukla,Vijay KR Trivedi, Munesh C Trivedic Encryption algorithm in cloud computing.
10. Educba https://www.educba.com/firewall-uses/

11. Fatsuma Jauroac, Haruna Chiromab, Abdulsalam Y.Gitalc, Mubarak Almutairid, Shafi'i M.Abdul hamid, Jemal H.Abawajy
12. https://en.wikipedia.org/wiki/Deep_learningCloud WAF: Overview and Benefits https://www.globaldots.com/blog/cloud-waf-overview-benefits
13. John R. Vacca and Scott R. Ellis Firewalls: Jumpstart for Network and Systems Administrators
14. Kumar Sahwal, ..., Kishore, ..., Singh Rathore, ..., & Moy Chatterjee, ... (2018). An Advance Approach of Looping Technique for Image Encryption Using in Commuted Concept of ECC. International Journal Of Recent Advances In Signal & Image Processing, 2(1)
15. Kumar, A., Chatterjee, J. M., & Díaz, V. G. (2020). A novel hybrid approach of svm combined with nlp and probabilistic neural network for email phishing. International Journal of Electrical and Computer Engineering, 10(1), 486.
16. N. Bhargava, S. Dayma, A. Kumar and P. Singh, "An approach for classification using simple CART algorithm in WEKA," 2017 11th International Conference on Intelligent Systems and Control (ISCO), Coimbatore, 2017, pp. 212-216, doi: 10.1109/ISCO.2017.7855983.
17. NaLu, Ying Yang: Application of evolutionary algorithm in performance optimization of embedded network firewall.
18. Naveen Kumar, Prakarti Triwedi, Pramod Singh Rathore, "An Adaptive Approach for image adaptive watermarking using Elliptical curve cryptography (ECC)", FirstInternational Conference on Information Technology and Knowledge Management pp. 89–92, ISSN 2300-5963 ACSIS, Vol. 14 DOI: 10.15439/2018KM19
19. P. Ravi Kumar, P. Herbert Raj, P. Jelciana Exploring Data Security Issues and Solutions in Cloud Computing.
20. Rathore, P.S., Chatterjee, J.M., Kumar, A. et al. Energy-efficient cluster head selection through relay approach for WSN. J Supercomput (2021). https://doi.org/10.1007/s11227-020-03593-4
21. Saurabh Singh, Young-Sik Jeong, Jong Hyuk Park A survey on cloud computing security: Issues, threats, and solutions.
22. Singh Rathore, P., Kumar, A., & Gracia-Diaz, V. (2020). A Holistic Methodology for Improved RFID Network Lifetime by Advanced Cluster Head Selection using Dragonfly Algorithm. International Journal Of Interactive Multimedia And Artificial Intelligence, 6 (Regular Issue), http://doi.org/10.9781/ijimai.2020.05.003

10

Smart Load Balancing in Cloud Using Deep Learning

Astha Parihar* and Shweta Sharma

MDS University, Ajmer, India

Abstract

Load adjusting in distributed computing is a significant administration and book-ing issue; in light of the fact that, in a distributed computing condition, every client may look for several virtual assets for running each undertaking. Consequently, choosing the best asset for playing out the errand is one of the most significant difficulties. Different burden adjusting calculations have been proposed to react to these issues; obviously, a portion of these calculations have some feeble focuses other than their solid focuses. Portable help figuring is another distributed com-puting model that gives different cloud administrations to versatile smart termi-nal clients through portable web get to. The nature of administration is a basic issue addressed by portable help figuring. In this paper, we show a progression of examination that concentrates on the best way to quicken the preparation of a disseminated AI Deep Learning model dependent on remote administration. Circulated Deep Learning has become the standard method of the present Deep Learning models being prepared. In conventional appropriated Deep Learning dependent on mass simultaneous equal, the impermanent log jam of any hub in the group will defer the estimation of different hubs in view of the successive event of coordinated hindrances, bringing about, generally speaking, execution debase-ment. Our paper proposes a heap adjusting methodology called Versatile Quick Reassignment (AdaptQR). In light of this, we fabricated a disseminated equal reg-istering model called the Versatile Powerful Simultaneous Equal (A-DIC). A-DIC utilizes a progressively loosened up coetaneous miniature to diminish the presen-tation utilization brought about by simultaneous activities while guaranteeing the consistency of the model. Simultaneously, A-DIC likewise actualizes the AdaptQR load adjusting methodology, which tends to the be the stray issue brought about by the exhibition contrast between hubs under the reason of guaranteeing the

Corresponding authors: asthaparihar9@gmail.com

Pramod Singh Rathore, Vishal Dutt, Rashmi Agrawal, Satya Murthy Sasubilli, and Srinivasa Rao Swarna (eds.) Deep Learning Approaches to Cloud Security, (145–166) © 2022 Scrivener Publishing LLC

precision of the design. The trials display that A-DIC can viably enhance preparation speed while guaranteeing the exactness of the model in the conveyed Deep Learning model preparation.

Keywords: Cloud computing (distributed computing), load balancing, mobile administration processing, cloud administration distributed AI, adaptive quick reassignment, A-DIC

10.1   Introduction

By rapidly progressing making and advancing the consumption and achievements of the web, this innovative arrangement has empowered another acceptance of recorded design known as distribution of computation, that is distributing computing. In this computing, the properties, for example CPU efficiency and efficient gadgets, are considered administrations that are rented and released by the consumer by means of the internet in a like-based behaviour. In distributed computing, a condition and the regular job of the specialist organisations are divided into two sections:

1. Suppliers of framework which deal with the remote working framework and rent the properties depending on the use of the valuing design.
2. Suppliers of the administrations who lease the assets from one or numerous foundation suppliers to serve the end clients.

By and large, distributed computing incorporates various disseminated servers offering varying types of assistance to the clients dependent on the solicitations in a versatile and solid system. The heap adjusting issue in distributed computing is another test, which, in every case, needs an appropriated arrangement. Since, practically speaking, it isn't generally conceivable and financially savvy to keep at least one specialist organization inert to react to required solicitations, along these lines, the undertakings cannot be relegated (assigned) to a suitable server. Burden adjusting alludes to the technique of reallocating the absolute burden to the exceptional hubs of an aggregate framework to viably use the assets, improve reaction time, and, at the same time, dispose of a status wherein a portion of the hubs are over-burdened although a few others are less stacked. The heap adjusting calculations are classified as static and dynamic calculations. Static calculations are, for the most part, reasonable for homogeneous and stable conditions and can yield entirely alluring outcomes in such situations. Be that as it may, they are typically unbendable and unfit to adjust with the dynamic changes of the highlights during runtime [1]. Then again, dynamic calculations are progressively adaptable and consider an assortment of highlights in the framework of the two modes, for example previously and during runtime. Burden adjusting is a procedure which improves the framework's exhibition by reallocating the heap among the processors in this manner and load adjusting in the cloud. Execution of the framework is one of the most significant difficulties of the specialists and analysts who endeavour in their research to examine the assets and issues (inconveniences) happening in running (executing) applications and the issue of booking the framework's presentation too. Reacting to the solicitations, overseeing, and planning with the end goal so that it yields the most ideal reaction in the briefest time requires preparation, booking, and the executives of disappointment. What's more, a significant number of the exploration discoveries, including hereditary calculation, subterranean insect state, computerized reasoning, and so forth, are utilized to take care of the issue of burden adjusting. In this paper, it is endeavoured to structure a keen framework for chiefs of the framework, which can give proposals to changing the framework's setup with the goal that they can upgrade the framework and improve heap adjusting through applying these changes.

10.2   Load Balancing

Burden adjusting is a reasonably new procedure that boosts systems and properties by giving a greatest throughput least reaction time. By splitting the traffic in the middle of the servers, details are able to send and be received instantly. Different computation is available that helps traffic stacked in the middle of available slaves. A fundamental instance of burden adjusting in our daily routine is the ability to recognise sites. Without load adjusting, clients could encounter delays, breaks, and conceivable long framework reactions. Burden adjusting arrangements, for the most part, apply excess servers that help a superior appropriation of the correspondence traffic with the goal that the site's accessibility is definitively settled. There are a wide range of sorts of burden adjusting calculations available, which can be classified predominantly into two groups. The following section will examine these two fundamental classifications of burden adjusting calculations [2].

The issue of burden adjusting in distributed computing is another test. Continuously, an appropriated arrangement is required. Since, practically speaking, it is not generally conceivable and financially savvy to keep at least one assistance supplier as inactiv to react to the required solicitations, the errands cannot be appointed (designated) to the fitting server, powerful burden adjusting independently for clients in the fog framework is extremely muddled, and the segments are available over a large area. Burden adjusting incorporates the methodology of reallocating the all-out burden to the remarkable hubs of an aggregate framework all together to viably use the assets, improve reaction time, and all the while evacuate details in which a portion of the hubs are over-burdened, as long as there is low pile up. Heap adjusting calculations are ordered as static and dynamic calculations. Static calculations are, for the most part, appropriate for homogeneous and stable conditions and can yield generally excellent outcomes in such conditions; in any case, they are typically unyielding and incapable of adjusting with dynamic transformation of the highlights in runtime. Then again, dynamic calculations are progressively adaptable. What's more, they bring into thought an assortment of highlight applicable errands in the framework in two modes during runtime. The procedure that is used to enhance the framework's presentation reallotment of the stack between processors is called load accommodation [3].

10.2.1   Static Algorithm

Static calculations partition traffic proportionally between servers. By this methodology, the traffic at the servers will be hated effectively. Arithmetic representation used to separate the traffic is correspondingly reported as a cooperative calculation. Notwithstanding, there were bunches of issues that showed up in this calculation. Along these lines, weighted cooperative effort was characterized to improve the basic difficulties related to cooperative effort. This representation shows that each slave has been allocated to a load and, as indicated by greatest elevated load, received many organisations. In this situation, all the loads are equivalent and each slave will get regular traffic.

10.2.2   Dynamic (Run-Time) Algorithms

Dynamic computation fixed exact loads on slaves and, via view of the whole system, a lite server was linked to adjust traffic. Be that as it may, choosing a proper server compulsory continuous correlation with the systems will prompt additional traffic included in the framework. In the examination between these two calculations, albeit cooperative calculations are dependent on straightforward principle, more loads are considered on servers and subsequently imbalanced traffic [4]. In any case, dynamic calculation predicated that an inquiry can be made much of the time on servers, but, once in a while, one rush hour gridlock will stall these questions being answered and, correspondingly, more included overhead can be recognized by the system.

10.3   Load Adjusting in Distributing Computing

Distributed computing is the up and coming age of figuring. Distributed computing innovation is developing quickly and the requests of customers are additionally developing quickly for additional administrations and better outcomes. IT enterprises are developing every day and with this development comes requirements for figuring and capacity assets. Information and data in enormous amounts are created and traded over a safe or unbound system, which further requires requirements for increasingly more propelled processing assets. To capitalize on their ventures, associations are opening their bases of activity to recently discovered virtualization advancements like Cloud processing. Presently, a day's kin can, without much of a stretch, have all that they need on the Cloud. Distributed computing gives assets in bounty to the customer on their interest. The requested assets might be programming assets or equipment assets. The models of distributed computing are conveyed equally and serve the necessities of various customers simultaneously [5].

There are different processors occupied with cloud engineering. Customers in a conveyed domain haphazardly produce a solicitation in any processor in the system. Its significant disadvantage is related to the task of assignments. The inconsistent task of the errand to the processor makes unevenness; a portion of the processors are vigorously over-burdened and some of them are underloaded. The principle goal of burden adjusting is to move the heap from over-burdened procedures to an underloaded procedure. Burden balancing is fundamental for productive activities in an appropriated situation. To accomplish better execution, least and quickest reaction time, and highest asset usage, we have to move the assignments between various hubs in the cloud arrangement [6]. The heap adjusting method is utilized to circulate assignments from over-burdened hubs to underloaded or sit hubs.

Cloud load adjusting is fundamentally the way toward appropriating or separating the remaining tasks at hand and diversifying processing assets across at least one accessible server. This sort of conveyance guarantees the most extreme throughput in a base reaction time. The remaining burden is separated among at least two servers, hard drives, arrange interfaces, or other diverse registering assets, which assist with empowering better asset use and improving framework reaction time. Along these lines, for a site with a high traffic rate, successful utilization of cloud load adjusting can guarantee better business progression. The regular goals of using load balancers are:

- To keep up structure solidness
- To improve framework execution
- To secure against framework disappointments

Cloud suppliers, such as Amazon Web Services (AWS), Microsoft Azure, and Google, offer cloud load adjusting to encourage simple dissemination of the remaining tasks at hand. For example, Amazon Web Services (AWS) offers a Flexible Load Adjusting (FLA) innovation to convey congestion between Flexible Compute Cloud (FCC) cases. The majority of Amazon Web Services' (AWS) controlled applications have Flexible Load Balancers (FLAs) introduced as the key building part. Thus, Azure's Traffic Manager allots its cloud servers' traffic over various server farms shown in Figure 10.1 [7].

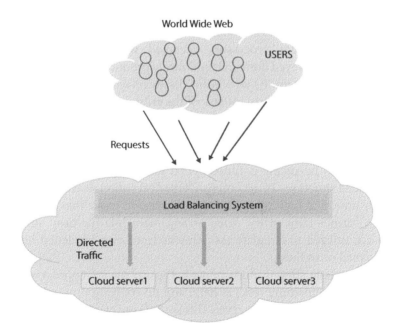

Figure 10.1 Cloud load balancing.

10.3.1   Working of Load Balancing

All things considered, load doesn't allude to just site traffic, but it similarly incorporates the CPU load and arranges the burden and memory limit of each server. A heap adjusting procedure consistently ensures that every single framework associated with the system has a similar measure of outstanding tasks at hand at any moment in time [8]. This guarantees that neither of them is unreasonably over-stacked nor under-used.

The stack balancer reliably passes on data depending on how clamouring each worker or centre is at a particular time. Beyond the stack balancer, the customer must postpone until his gadget is readied, which is probably unreasonably tiring and demotivating for him and it is not always proposed every time. Various forms of facts, like occupations maintaining up in the line, CPU's taking care of rate, action appearance rate, and so forth, are exchanged among the processors during the stack converting manner. Frustration with the right utilization of the stack balancers can initiate enormous issues and statistics getting lost is considered one of them.

Different groups or systems make use of many load balancers and numerous stacks accommodate calculations like static and dynamic load accommodation. One of the greatest ordinarily used strategies is Round-Robin Load Adjusting.

It raises the client's request to particular related slaves. The considerable benefit is its modesty of execution. The stack balancers analyze the structure pules during defined time spans to ensure whether each hub is performing well or not shown in Figure 10.2.

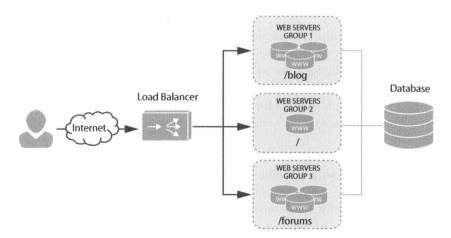

Figure 10.2 Working of load balancer.

Conventional burden adjusting arrangements exclusively depend on if the restrictive equipment is essentially accessible or arranged in a server farm and if it requires a group of a profoundly refined IT work force to constantly introduce, tune, and keep up the framework. Huge organizations with extraordinary IT financial plans can understand the advantages of improved execution and unwavering quality. In the time of distributed computing, equipment-based arrangements have genuine downsides; they can't bolster cloud load adjusting in light of the fact that cloud framework merchants regularly do not permit the client or exclusive equipment inside their workplace [9].

Fortuitously, programming based burden balancers can send the presentation and resolute fine benefits of system based preparations at a much better value. Since they run on moderate and simple to acquire equipment, they are reasonable in any event for smaller organizations. Programming based burden balancers are perfect for cloud load adjusting as they can, without much of a stretch, bundles.

10.4   Cloud Load Balancing Criteria (Measures)

Different boundaries are considered in cloud load adjusting methods to show various kinds of burden adjusting shown in Figure 10.3.

Important Overhead: When a balance calculation is performed, it finalizes the estimation of added overhead; added overhead indicates the enhanced errors and the connection between processors and procedures.

Transactional Force is used for ascertaining the ended errands. This is really excessive for enhancing the frameworks' proficiency.

Effectiveness is utilized for controlling the framework's performance. This degree must improve the realistic prices. For instance, it ought to diminish the challenge time whilst keeping the postponements at a first-rate degree.

Property Misuse is utilized for controlling the exploitation of the property. This is an area that needs to be progressed for talented burden adjusting.

Versatility is the capacity of a calculation to perform load offsetting for a framework with a limitless range of hubs. This model needs to be advanced.

Reaction Time is the time considered for reacting that has been circulated inside the framework by way of a selected burden adjusting calculation. This boundary should be restricted.

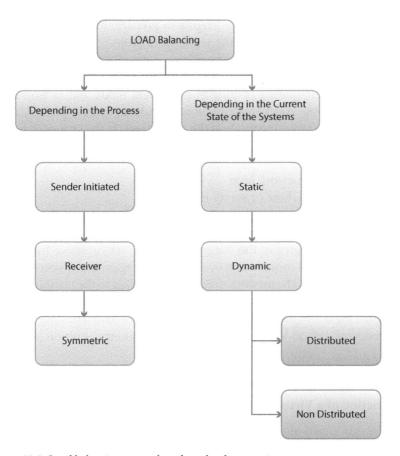

Figure 10.3 Load balancing system based on cloud computing.

Fault Resilience: is the capacity of a calculation to dully execute load adjusting if there is an occurrence of association disappointment. Burden adjusting ought to be a decent mistake resistance method.

Movement Time is the hour of relocation of errands or assets starting with one hub then onto the next. This time must be limited to expand the framework's productivity.

10.5   Load Balancing Proposed for Cloud Computing

In distributed computing, at anything point, a digital gadget (VM) is stacked with numerous errands and those undertakings ought to be distributed and dispatched to much less stacked VMs in a similar server farm and it should not be excessively lengthy to prepare. Inside the VM stage, load adjusting is performed in the server farm [10].

Past usage of burden adjusting has been done in a programming mode and as nearby Machine learning. In the proposed technique, load adjusting for distributed computing is done in equipment mode and as worldwide streadeep learning. As expressed before, in past strategies, load adjusting was done in programming mode at the server level. In the proposed framework, load adjusting is performed at two levels:

1. Entire framework level
2. Virtual machine or server level.

At both tiers, load adjusting is determined so that if there is no heap adjusting at any degree of the framework, the server or whole framework that is less stacked is moved to an over-burdened region in the system mode. Within the proposed framework, paying little thought to the planning calculations, the progressions applied to the framework result in heap adjusting; alongside these strains, the proposed calculation for load adjusting capabilities performs well on VMs in the dispensed computing condition [11].

10.5.1   Calculation of Load Balancing in the Whole System

Restriction of a Virtual System

$$C_j = Pe_{nj} * Pe_{mipsj} + VM_{bwj,}$$ (10.1)

in which Pe_{nj} is the variety of processors in a VM_j, Pe_{mipsj} j is the million instructions in step with 2d of all of the processors in i VM_j and VM_{bwj} j is the ability of the VM_j communication bandwidth.

Capacity of All the VMs

$$C = \sum_{i=1}^{m} C_j$$ (10.2)

Capacity of the statistics middle is the whole potential of all of the virtual machines. Load in a VM is the complete length of the action allocated to a VM.

$$L_{VMj}t = N (T, t)/S (VM_j, t)$$ (10.3)

The load of VM may be calculated as the variety of the obligations at the time (t) inside the provider queue of vmi divided by means of the carrier charge of vmi at the time (t). The load of all the virtual machines in a records middle is calculated as follows:

$$L = \sum_{i=1}^{m} L_{VMj} \tag{10.4}$$

Load Standard Deviation

$$\sigma = \sqrt{\frac{1}{m}\sum_{i=1}^{m}(PT_i - PT)^2} \tag{10.5}$$

Inside the wake of finding the extremely task handy, same old deviation, the framework had to pick whether or not burden adjusting had to be completed or not. For this reason, there are two capacity states: (1) seeing whether or not the framework is adjusted and (2) discovering whether the whole framework is immersed or not (regardless of whether or not the entire accumulation is over-burdened or not). In the case that it is extremely over-burdened, load adjusting is insane.

If there are VM's gathering and the standard deviation of the VM load (σ) is not always precise or equal to the restricted set of the conditions (Ts) [1-0], at that factor, the framework is adjusted or else it is in an insecure state.

If there are over-burdened bunches, the best thing to do is hand off the digital system bunch; if the bunch surpasses the finest limit of the collection, the gathering is over-burdened.

10.6   Load Balancing in Next Generation Cloud Computing

Burden adjusting can be accomplished by better asset usage and improving the general execution of the framework. The up and coming age of cloud dynamics includes multiservice ideas in which each server group exclusively handles an extraordinary mixed media task and each customer demands an exceptional sight and sound transporter at a particular time. To accomplish effective burden conveyance, a heap balancer is utilized wherein errands are retrieved from various areas and afterwards disseminated to the server farm.

Development within the cloud discipline aims to efficiently develop sensors and devices, bringing individuals, gadgets, and related registering nearer. The Internet of Things (IoT) is a cutting edge innovation that allows billions of shrewd items to chat with others while a human carries on and is considerably progressively advantageous. The IoT depends on a remote sensor network (WSN) and Zigbee is one of the most celebrated WSN conventions. Nonetheless, Zigbee's AODV directing stack does not have the load strength instrument to oversee bursts in traffic [12].

Information preparing with load adjusting incorporates utilizing sensors implanted into the framework, which incorporated transportation, correspondence, structures, social insurance, and different utilities and sensors on client gadgets and wearable and home hardware have brought about the up and coming Internet of Things (IoT) [13]. A burden adjusting cloud with the IoT improves precision and productivity and decreases human mediation to be viable for getting information.

Worldwide burden adjusting innovation clears up inconveniences of burden adjusting in one zone and in different groups between selective models. These are basically reflected server bunch hubs with indistinguishable substances and can stay away from certainty blockage presented by high simultaneous access in sites.

The fundamental point of the burden adjusting issue on the distributed computing stage is to productively plan setting VMs. Nature enlivened unique burden adjusting stays a difficult worldwide streadeep Learning issue because of the following issues: heterogeneous structure of the cloud, registering assets, and the nature of administration (QoS) orders by applications. The significant inspiration driving understanding heap offsetting techniques in the cloud with rising advances are as follows which can be shown in Figure 10.4:

- To plan a framework model for task assignment in distributed computing with developing advancements
- To create nature roused dynamic burden adjusting

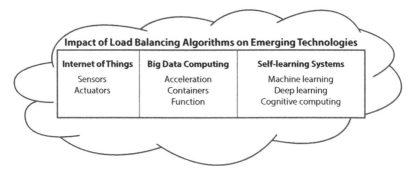

Figure 10.4 Impact of load balancing.

10.7 Dispersed AI Load Adjusting Methodology in Distributed Computing Administrations

Versatile help processing is a fundamental supporting innovation for portable web and distributed computing and its development adds dynamic and wise abilities to disseminated figuring. In the portable distributed computing mode, the gigantic measure of data handling and monstrous information stockpiling that at first acted in brilliant versatile terminals are moved to the "cloud", in this manner decreasing product and equipment necessities of the savvy portable terminals. In any case, portable help registering faces some new difficulties, the most significant of which is to guarantee the accessibility and unwavering quality of cloud administration. Both the scholarly community and industry have put colossal endeavours into how to balance out the nature of cloud administration, decrease deferral, and improve the nature of experience for endorsers and have gained noteworthy ground as of late [14].

Cloud-based Deep Learning calculations, for the most part, embrace equal and disseminated usage to guarantee the exhibition of QoS. The iterative-merged calculation is a fundamental subset of Deep Learning calculations. This calculation will initially create an arbitrary arrangement and, by iterative preparing of the information to intermingling, it will be persistently enhanced to accomplish a large wellness arrangement. The iterative-joined calculation generally cuts the information right off the bat and afterward utilizes the QIP model for disseminated model preparing.

The principle execution issue of the QIP model is the stray issue. The stray issue occurs due to a lopsided group load and the terrible showing of a processing hub hinders the general running velocity. The presence of a simultaneity design makes the time amount of each emphasis of preparing reliant upon the most noticeably awful performing registering hubs, which hinders the pace of other processing hubs. As the equal size of figuring models builds, the stray issue turns out to be progressively obvious. Since the time cost of the figuring hub for every cycle of preparing changes powerfully with the quantity of registering hubs expanding, the likelihood of bunch load unevenness in every emphasis increases. Thus, the presentation of model preparing was indicated as diminished in the QIP.

Registering hubs in the cloud condition by and large have distinctive executions, so the stray issue in circulated Deep Learning dependent on cloud administration is especially fundamental. To guarantee the accessibility and unwavering quality of cloud administration, an increasingly appropriate registering model is deprived to address the stray issue [15].

10.7.1   Quantum Isochronous Parallel

For dispersed preparing, Deep Learning commonly utilizes the QIP model, that is, Spark and Graph. The registering process includes a progression of supersteps of duration (T) isolated by a coetaneous obstruction. A progression of activities in two simultaneous obstruction spans is utilized by supersteps. In every step, all registering hubs perform iterative computation simultaneously. In each superstep, all registering hubs perform iterative computations simultaneously.

Subsequent to computing the current emphasis, every hub transfers the refreshed nearby boundaries to the boundary slave and, afterward, put in the synchronization hindrance and pauses. As all registering hubs complete the count and submit the procedures, the boundary server refreshes the worldwide model boundaries and points them all to figuring hubs; at that point, the synchronization obstructions are discharged with the goal that all processing hubs go to the following superstep together. This synchronization system permits an equal Deep Learning calculation of the QIP model to be serialized, guaranteeing worldwide consistency of boundary refreshes and correct calculation execution [16].

The time cost analysis for QIP is defined by following formula:

$$S_{uperset} = \max \{w1/r1, w2/r2, wn/rn\} + \max\{x1, x2, xn\}^*K+L \tag{10.6}$$

$T_{superstep}$ suggests the time value compulsory for super steps in which x_i shows the count of CPU cycles expected for the present iteration of working. i and ri, describe the process speed of the CPU, ai describes the box size delivered or received through a process, I and k represent the reverse of the bandwidth, and L shows the shallow time of simultaneous stumble.

Collision time is the time it takes for a computing node with good performance to wait for a computing node with poor performance

The collision time of each node is:

$$\nabla \text{Max } (t1, t2, ti)\text{-}t_i \tag{10.7}$$

QIP offers a basic and compelling model for the parallelization of iterative-focalized calculation. Figure 10.5 indicates the processing model of QIP. Be that as it may, it very well may be visible from Equation 10.7 that there is a huge stray issue within the QIP. Each superstep inside the figuring system is performed by way of the coaching pace of the slowest hub, with visit utilization of the synchronization boundary, which means that

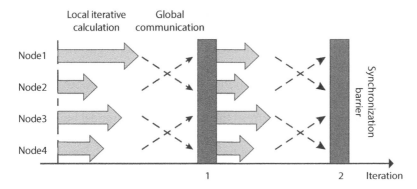

Figure 10.5 Quantum isochronous parallel computing model.

the powerful execution of QIP has exacting stipulations for load adjusting. Resulting from a choppy organization load, the presentation of the QIP processing model might be essentially dwindled. For instance, research shows that when the LDA subject model suddenly spikes in demand for 32 machines under QIP, the synchronization boundaries can be up to multiple times longer than the cycles. Nonetheless, research shows that even on account of bunch load adjusting, the arbitrariness of execution time that every hub takes for emphasis in the preparation procedure will likewise create a deference in different hubs [17].

The registering procedure is executed by the preparation momentum of the slowest hub, with visit utilization of the synchronization hindrance that is the proficient performance of QIP with exacting necessities for load adjusting. On account of a lopsided bunch load, the exhibition of the QIP figuring model will be altogether diminished. Anyway, even on account of group load adjusting, the irregularity of implementation time that every hub consumes for emphasis in the preparation procedure will likewise postpone different hubs [18].

10.7.2   Phase Isochronous Parallel

To allocate the stray issue in QIP, a Phase Isochronous parallel processing model is proposed. PIP utilizes the adaptation to internal failure of the iterative-focalized calculation to structure a progressively adaptable consistency synchronization technique. Accordingly, proficiently facilitating the stray problem is commonly circulated in Deep Learning systems dependent on PIP.

PIP permits each processing hub to play out different emphases of preparation utilizing a nearby model rather than a worldwide model. In any case, when the quantity of cycles of some other registering hub surpasses the slowest hub s times (where s is characterized as the staleness limit), the synchronization hindrance will compel the processing hub to sit tight for the slowest hub to finish the emphasis. Worldwide model boundaries are then gathered to refresh the neighbourhood model boundaries.

The recurrence of events of the synchronization hindrance is diminished, lessening the time cost of preparing the PIP model. Simultaneously, by setting the staleness edge (s), we can ensure that the neighbourhood model duplicate (M) contains the updates of every single other hub during the cycle times in [0,C-s-1] and pieces of the updates during the emphasis times in [C-s, C?s-1], where C speaks to the current cycle number, consequently guaranteeing the intermingling of equal execution of Deep Learning calculations shown in Figure 10.6 and Figure 10.7 [19].

This examination shows that PIP can adequately lighten the stray issue, however, despite everything, it has the following issues:

a. In the test condition with comparative group hub execution, the PIP will be excessively slack because of the synchronization system, which will defer the synchronization of the hubs. Thus, the model update is seriously obsolete and the combination speed and final precision of the model are altogether diminished.

b. The invariance of the staleness limit (s) makes the PIP incapable of adapting well to ongoing changes of hub execution during the model preparing process.

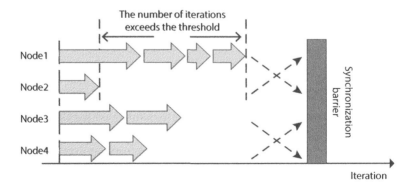

Figure 10.6 Phase isochronous parallel model.

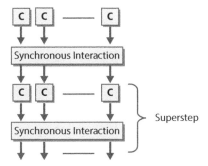

Figure 10.7 Dynamic isochronous coordinate strategies.

10.7.3 Dynamic Isochronous Coordinate Strategy

To take care of the two issues referenced above, the dynamic simultaneous equal model is proposed. In view of the PIP, DIC has made pointed enhancements. For the issue of PIP not be applicable for comparative execution of the hub, the DIC proposes the idea of the feeble edge (w), which specifies that when the cycle number of the slowest registering hub arrives at w every emphasis, the boundary is compelled to be refreshed in synchronization. This instrument viably controls the obsolete level of model boundary updates among hubs and, in any event, when the presentation of every hub is exceptionally close, the synchronization condition can be reached generally rapidly. For the issue of the invariance of the staleness edge (s) in the PIP, the DIC performs constant execution checking on each processing hub by including the exhibition observing module and, afterward, powerfully altering the staleness edge (ds) as indicated by the presentation distinction of each figuring hub. In this manner, it can adequately adjust to the constant changes of hub execution because of wild factors in the model preparing process, for example, outside impedance.

10.8 Adaptive-Dynamic Synchronous Coordinate Strategy

In this segment, the limitations of DIC when tackling the stray issue are presented in detail from the outset. At that point, the AdaptQR load-adjusting methodology and A-DIC equal registering model are proposed and, based on Caffe, a disseminated Deep Learning system is executed with the possibility of A-DIC and a parametric server, which viably tackles the issues in DIC. At last, the rightness of A-DIC is investigated hypothetically shown in Figure 10.8 and Figure 10.9.

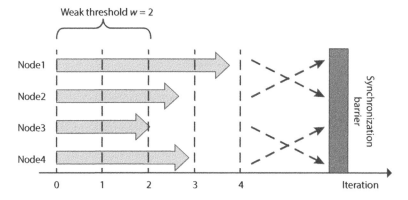

Figure 10.8 Weak threshold procedures in DIC model.

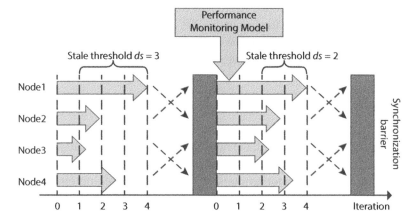

Figure 10.9 DIC model staleness portal dynamic adjustment procedure.

10.8.1 Adaptive Quick Reassignment (AdaptQR)

To successfully address the stray issue brought about by the profoundly unequal group load in circulated Deep Learning model preparing, this paper proposes a heap adjusting procedure in light of the remaining task at hand for dynamic redistribution: AdaptQR. AdaptQR can decrease the heap contrasts between hubs by viably altering the outstanding task at the hands of every hub in each cycle, as indicated by the diverse registration of exhibitions in the group.

In the preparation of a dispersed Deep Learning model dependent on iterative-merged calculation, each registering hub will more than once enter the following iterative preparation before arriving at the stop condition. In the conventional preparation of a disseminated Deep Learning model, informational collection for every emphasis of each figuring hub is an equivalent fixed size. AdaptQR doles out fewer calculations to moderate hubs, while distributing more calculations to quick hubs by redistributing the measure of calculations rapidly in every emphasis of each processing hub, adjusting the hour of finishing a solitary cycle among hubs subsequently by implication adjusting the heap of the bunch and improving the exhibition of model preparing [20].

10.8.2   A-DIC (Adaptive-Dynamic Synchronous Parallel)

Caffe is a profound studying structure known for its quick preparation and simple figure definition. Nonetheless, the open source adaptation of Caffe does not bolster conveyed Deep Learning. In this paper, we actualize an appropriated Deep Learning structure dependent on Caffe utilizing the possibility of an A-DIC and parametric server. The structure is appears in Figure 10.10.

The boundary server, for the most part, incorporates the following modules: the worldwide boundary, the board module, the bunch execution checking module, the dynamic simultaneous control module, and the remaining burden redistribution module.

The worldwide boundary is the executive's module and is utilized to keep up the worldwide model boundaries. It speaks with the registering hubs by keeping up Pthreads and bosses the presentation of the figuring hubs and the quantity of emphases per figuring process. The group execution checking module is responsible for gathering and dealing with ongoing exhibition information transferred by all registering hubs. In the wake of breaking down the presentation of each processing hub dependent on their systems, the dynamic coordinated control module and the remaining burden redistribution module correct the feeble limit (w), the staleness edge (s), and the redistributed outstanding task at hand (m I).

Figuring hubs and computing hubs chiefly incorporate the following modules: the information stockpiling module, the processing module, the staleness synchronization module, and the exhibition checking module.

The statistics stockpiling module is applied to keep the preparation subunits of each figuring hub. As indicated via the assigned incredible mission to hand m i, the processing hub peruses the reference scale data from the data board module for preparation. The registering module accommodates numerous figuring paperwork, which might be utilized to check nearby fashions. The staleness synchronization module will powerfully alter the requirement states of the synchronization boundary as indicated by using the contemporary frail facet (w) and the staleness part (s). On the off chance that the situations are met, the registering module may know enough to continue with its

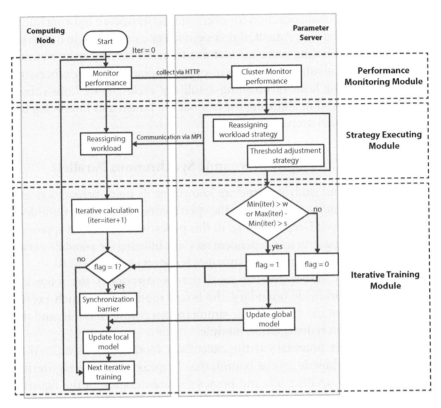

Figure 10.10 A-DIC training block diagram.

replacement simultaneously after the modern-day iteration. Anyhow, the processing module could be told to refresh locally and enter the subsequent cycle. The exhibition looking at the module is applied to screen the contemporary execution guidelines of processing hubs, for instance, reminiscence, CPU, i/o, to arrange and send them to the boundary server for examination.

The MPICH finishes the correspondence between the processing hub and the boundary slave. The registering procedure of the processing module delivers a refreshed slope by no concurrent transfer and the worldwide boundaries are sent to the figuring hub by Pthreads. This correspondence mode can successfully lessen the danger of correspondence blocking [21].

10.9 Conclusion

To absolutely deal with the stray difficulty delivered through different versions in hub execution as the practice of a scattered deep mastering design, this journal introduces a piece load run-time redistribution manner (adapter) that could effectively decrease the heap contrasts between hubs through changing the top notch mission at the hand of each hub in every cycle as indicated via the numerous registering exhibitions in the bunch. Then, primarily based on DIC, we actualize any other identical registering design and A-DIC through an incorporating adapter. Above bankruptcy, we exhibit that A-DIC can viably adapt to the run-time fluctuation of hub execution in proper conditions and suggests the stray issue totally through trials. In any case, the exactness and union velocity of A-DIC is notable.

References

1. Abhishek Kumar & Jyotir Moy Chatterjee & Pramod Singh Rathore, 2020. "Smartphone Confrontational Applications and Security Issues," International Journal of Risk and Contingency Management (IJRCM), IGI Global, vol. 9(2), pages 1-18, April.

2. Alam M, Khan ZA (2017) Issues and challenges of load balancing algorithm in cloud computing environment. Indian J Sci Technol. https://doi.org/10.17485/ijst/2017/v10i25/105688

3. Arani M, Souri A, Rahmanian A (2019) Resource management approaches in fog computing: a comprehensive review. J Grid Computer. https://doi.org/10.1007/s10723-019-09491-1

4. Baek J, Kaddoum G, Garg S, Kaur K, Gravel V (2019) Managing fog networks using reinforcement learning based load balancing algorithm. IEEE Wirel Commun Netw Conf. https://doi.org/10.1109/WCNC.2019.8885745

5. Bhargava, N., Bhargava, R., Rathore, P. S., & Kumar, A. (2020). Texture Recognition Using Gabor Filter for Extracting Feature Vectors With the Regression Mining Algorithm. International Journal of Risk and Contingency Management (IJRCM), 9(3), 31-44. doi:10.4018/IJRCM.2020070103

6. Chilimbi, T. M., Suzue, Y., Apacible, J., & Kalyanaraman, K. (2014) Project adam: Building an efficient and scalable deep learning training system. In OSDI (Vol. 14, pp. 571-582).

7. Cun, Y. L., Boser, B., Denker, J. S., Henderson, D., Howard, R. E., Hubbard, W., et al. (1990). Handwritten digit recognition with a back-propagation network. Advances in Neural Information Processing Systems, 2(2), 396-404.

8. Dai, W., Kumar, A., Wei, J., Ho, Q., Gibson, G., & Xing, E. P. (2014). High-performance distributed ML at scale through parameter server consistency models. In National conference on artificial intelligence (pp. 79-87).

9. Deng, L. (2012). The MNIST database of handwritten digit images for machine learning research [Best of the Web]. IEEE Signal Processing Magazine, 29(6), 141-142.

10. Jia, Y., Shelhamer, E., Donahue, J., Karayev, S., Long, J., Girshick, R., *et al.* (2014) Caffe: Convolutional architecture for fast feature embedding. In *Proceedings of the 22nd ACM international conference on Multimedia* (pp. 675–678). ACM.

11. Kumar, A., Chatterjee, J. M., & Díaz, V. G. (2020). A novel hybrid approach of svm combined with nlp and probabilistic neural network for email phishing. International Journal of Electrical and Computer Engineering, 10(1), 486.

12. Li, M., Zhou, L., Yang, Z., Li, A., Xia, F., Andersen, D. G., *et al.* (2013) Parameter server for distributed machine learning. In *Big learning NIPS workshop* (Vol. 6, p. 2).

13. N. Bhargava, S. Dayma, A. Kumar and P. Singh, "An approach for classification using simple CART algorithm in WEKA," *2017 11th International Conference on Intelligent Systems and Control (ISCO)*, Coimbatore, 2017, pp. 212-216, doi: 10.1109/ISCO.2017.7855983.

14. Naveen Kumar, Prakarti Triwedi, Pramod Singh Rathore, "An Adaptive Approach for image adaptive watermarking using Elliptical curve cryptography (ECC)", First International Conference on Information Technology and Knowledge Management pp. 89–92, ISSN 2300-5963 ACSIS, Vol. 14 DOI: 10.15439/2018KM19

15. Naveen Kumar, Prakarti Triwedi, Pramod Singh Rathore, "An Adaptive Approach for image adaptive watermarking using Elliptical curve cryptography (ECC)", First International Conference on Information Technology and Knowledge Management pp. 89–92, ISSN 2300-5963 ACSIS, Vol. 14 DOI: 10.15439/2018KM19

17. Rathore, P.S., Chatterjee, J.M., Kumar, A. *et al.* Energy-efficient cluster head selection through relay approach for WSN. J Supercomput (2021). https://doi.org/10.1007/s11227-020-03593-4

18. *Singh Rathore, P., Kumar, A., & Gracia-Diaz, V. (2020). A Holistic Methodology for Improved RFID Network Lifetime by Advanced Cluster Head Selection using Dragonfly Algorithm. International Journal Of Interactive Multimedia And Artificial Intelligence, 6 (Regular Issue), 8.* http://doi.org/10.9781/ijimai.2020.05.003

19. Xing, E., Ho, Q., Dai, W., Kim, J. K., Wei, J., Lee, S., *et al.* (2015). Petuum: A new platform for distributed machine learning on big data. IEEE Transactions on Big Data, 1(2), 49–67.

20. Zhang, J., Sha, C., Wu, Y., Jian, W., Li, Z., Ren, Y., *et al.* (2016). The novel implicit LU-SGS parallel iterative method based on the diffusion equation of nuclear reactor on GPU cluster. Computer Physics Communications, 211, S0010465516301965.

21. Zhang, J., Xiao, J., Wan, J., Yang, J., Ren, Y., Si, H., *et al.* (2017). A parallel strategy for convolutional neural network based on heterogeneous cluster for mobile information system. Mobile Information Systems, 2017, 3824765. https://doi.org/10.1155/2017/3824765\

11

Biometric Identification for Advanced Cloud Security

Yojna khandelwal[1]* and Kapil Chauhan[2]

[1]MDS University, Ajmer, India
[2]Aryabhatta Engineering College, Ajmer, Rajasthan, India

Abstract

Biometric discovery has become increasingly popular in recent years. Biometric identification is a trusted and efficient way of recognizing users. The on ground practical implementation of protection from the loss, misuse, or theft of biometric data needs an efficient privacy preserving method. Currently, only the house primitive cryptographic algorithms are being used in identification and authorization methods and old existing techniques are prone to many security threats. With the initiation and development of the cloud computing mechanism, database owners are given a chance to outsource large amounts of biometric data and identification tasks to the cloud to avoid the heavy expenses spent in storage and computation costs, which in turn open various channels of data security. This biometric dataset is then encrypted using cryptographic algorithms and deployed to the cloud servers. To recognize biometric data, the database owner authorizes by generating a credential for a user and then, the biometric data is submitted to the cloud server. Deep Learning and Big Data are the two very emerging technologies. The large amount of data gathered by organizations is used for many purposes, like for resolving problems in marketing, medical science, technology, national intelligence, etc. In this current world, old house data processing units are not very efficient to handle, process, or analyze because the collected data is unstructured and very complex. Because of this, Deep Learning algorithms which are fast and efficient in solving the backlogs of traditional algorithms are in use now a days. Biometrics use the pattern recognition technology of Digital Image Processing for identifying unique features in humans. The most commonly applied or considered biometric modalities are fingerprint impression, facial landmarks, iris anatomy, speech recognition, handwriting detection, hand geometry recognition, Finger

*Corresponding author: yojnakh22@gmail.com

Pramod Singh Rathore, Vishal Dutt, Rashmi Agrawal, Satya Murthy Sasubilli, and Srinivasa Rao Swarna (eds.) Deep Learning Approaches to Cloud Security, (167–188) © 2022 Scrivener Publishing LLC

vein detection, and signature identification. Biometric technology is applied in various fields, security systems being one among them. Biometrics can be effectively applied to eliminate one of the cloud's security issues--identity theft. A wide range of biometric authentication system protocols and implementations in the cloud environment, especially to combat identity theft, have been proposed.

Keywords: Biometrics, deep learning, cryptographic, security systems, protocols,

11.1   Introduction

11.1.1   Biometric Identification

Biometrics is the process of peruse physical or behavioral traits related to any user in order to establish their identity.

Biometrics allow a user to be recognized or authorized depending on a dataset of identifiable and authorizable data which are user related and particularly specific to a user.

Biometric Authentication is the process of making the comparison of the human biological features data and making a resemblance to the recorded feature set to provide the authentication shown in Figure 11.1 [1].

- In the reference model, the first one is to put away in a database or a protected versatile component like a savvy card.
- The information put away is then contrasted with the individual's biometric information to be validated. Here, it is the individual's character that is being checked.
- Biometric distinguishing proof includes the deciding characteristics of humans.
- The point is to catch biometric information from this individual. It tends to be a photograph of their face, a record of their voice, or a picture of their unique mark.
- This information is then contrasted with the biometric information of a few different people kept in a database.
- In this mode, the input is a usual one: "Who are you?"

The most used use cases of biometric technologies:

1. Border, travel, and movement control
2. Law enforcement and open security
3. Civil ID for resident, occupant, voter, or identification proof
4. Military (foe or partner recognizable proof)

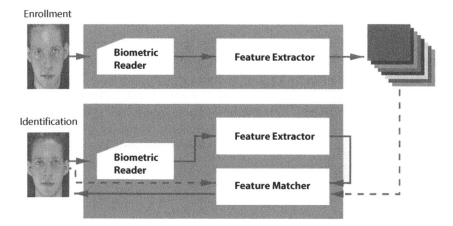

Figure 11.1 Biometric identification.

5. Healthcare and appropriations
6. Physical and coherent access
7. Commercial applications by purchaser or client recognizable proof

11.1.2   Biometric Characteristic

Ideal biometric data should have the following requirements:

Universality: each person should have the feature
Uniqueness: no two persons should match

11.1.3   Types of Biometric Data

11.1.3.1   Face Recognition

Face identification can be done in various ways. The major process in the recognition system is capturing an image in the spectrum that is visible and using a camera that is easily affordable or using the infrared patterns of the facial heat emissions of the user. The facial recognition method typically takes the key features of the human face in the light and models them into the given algorithm to store the data of the eyes, nose, ears, etc. These features are taken in consideration to the fact that these do not change much over time. Principal Component Analysis and Local Feature Analysis are some traditional methods used in this technology and now neural networks have come into existence for making facial recognition as fast as anything [2].

11.1.3.2   Hand Vein

Hand veins provide an easy way to identify users using personal identification marks. The images of hand veins are captured using infrared cameras. Patterns on the hand veins are unique in the humans and are very difficult to change, thus bringing in a good biometric identification method in order to provide security. This method attains recognizable identification accuracy and is normally accepted by the people. However, till now there was no marked presence of a hand vein based biometric system that was able to show a superior capability in making automatic personal identification. It is very difficult to attain the highest level of security in the hand vein system, like the hand geometry system, but it is still is a good mechanism.

11.1.3.3   Signature Verification

This security mechanism uses the analysis of a hand signature using a dynamic mechanism to authenticate the users. This methodology is based on the pressure, speed, and angle of the signature made by the user. This security mechanism is used in many areas such as e-business applications and other applications where the method of authentication includes the signature matching mechanism as well [3].

11.1.3.4   Iris Recognition

Iris recognition technique uses the iris of the human ey in order to provide authentication. The iris is the colored area around the pupil which is unique for every individual. The video based image acquisition technology is used to capture the iris details of the human beings. This technology has been in use for years, but now a days, there is a retardation of the use of the technology. This mechanism is used in identification as well as the verification of the identity of the human beings. The advancement in the technology has been so good that recognition can be done even through eye-glasses and contact lenses. The main advantage of this technology is that there is no physical contact with the scanner device. This technology is being used for every human irrespective to ethnic groups and nationalities [4].

11.1.3.5   Voice Recognition

Voice recognition technology has a history of development from the last few years. The acoustic features of the voice are in use for this recognition technology, as it differs in every individual.

The acoustic feature of the voice consists of all the required parameters for making a recognition, such as size, shape of the throat, anatomy features that affect the voice and voice pitch, and speaking style, which comes under learning behavioral patterns. The incorporation of the patterns is learned by the algorithm and then voiceprints are developed using them, which are in turn used for recognition. Text dependent, text prompted, and text independent are three types of spoken styles which the voice recognition method employs. A text dependent input method is deployed by most of the voice recognition methods, as this method involves the enrollment and selection of one or many voice passwords. A test prompted input is used in the case of imposters [5].

The various technologies considered in the process and storage of voiceprints are matrix representation, Hidden Markov models, neural networks, pattern matching algorithms, and decision trees. Behavioral attribute changes lead to performance degradation of the voice and from enrollment using one telephone switching for verification to another telephone. The main problem here is to solve the issues to voice changes by aging or any other short term throat problems. Voice recognition engines are developed and marketed by many companies, often as part of a large voice control processing and switching system. This technology implementation needs additional hardware by using existing voice-transmissio technology and microphones and allowing recognition over long distances through ordinary telephones (wired or wireless) [6].

11.1.3.6   Fingerprints

Now a days, the most familiar, used, advanced, and easily accessible recognition technology is fingerprint matching technology. The fingerprint pattern, the friction ridges, and valleys of anyone's fingerprint is unique for that person only. It cannot be reproduced. For decades now, in many countries, law enforcement has identified and determined user identity by mapping the key points of fingers' ridge endings. Even for identical twins, the fingerprints are different. This recognition technology is one of the world's most commercially available and used biometric technologies. Now for desktops and laptops as well, fingerprint recognition devices are widely available. Now, users do not need to type passwords; instead, only a touch can provide them instant access. This recognition system can also be used in identification mode [7].

The light and strong patterns which are on the tip of the finger are considered fingerprints and have been used excessively for personal purposes of people. Fingerprint formation biological properties are well defined and fingerprints have been in use for identification purposes for centuries now. From the beginning of the 20th century, fingerprints were used extensively for identification and recognition of criminals by many forensic departments around the globe. The fingerprint-based biometric system offers positive authentication with a very high level of accuracy, and tiny solid state sensors to read fingerprints can be easily placed in various systems such as mobile phones, etc. Fingerprint-based identification technology is becoming very popular in a number of regular and commercial applications, such as cellular phone access, welfare disbursement, and laptop computer log-in. The availability of small, tiny, compact, and cheap solid state scanners, as well as robust fingerprint matching methods and algorithms, are two important factors in making fingerprint-based identification systems popular worldwide. In current times, the highly used mechanism for security is based on fingerprints only [8].

11.2   Literature Survey

Pawle and Pawar [8] proposed a simple and easy face identification method which checks username and face image with the stored data. In addition to that, Siregar *et al.* proposed face recognition implementation and application of the Eigen face and REST concepts, respectively. Kisku and R showcased a system of face recognition system using Principal Component Analysis decided stances in which invariant points of SIFT are reduced shown in Figure 11.2 and Figure 11.3 for modalities measurements.

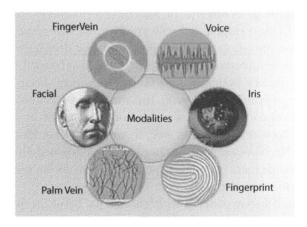

Figure 11.2 Biometric modalities.

Biometric Technology	Accuracy	Cost	Devices required	Social acceptability
ADN	High	High	Test equipment	Low
Iris recognition	High	High	Camera	Medium-low
Retinal Scan	High	High	Camera	Low
Facial recognition	Medium-low	Medium	Camera	High
Voice recognition	Medium	Medium	Microphone, telephone	High
Hand Geometry	Medium-low	Low	Scanner	High
Fingerprint	High	Medium	Scanner	Medium
Signature recognition	Low	Medium	Optic pen. touch panel	High

Figure 11.3 Biometric modalities measurement.

Rathi and Dubey [9] proposed a system using hand contour matching in which the Euclidian distance was used to measure the various distances in hands from specific points. Along with traditional methods, Yassin *et al.* [20] proposed a method with merits such as secure password changeability, mutual authentication, revocation, and user anonymity.

In this paper, authentication protocol using quantum identity has been put forth. It is based on optical transmission and facial feature recognition.

A secure kNN computation scheme on encrypted databases was introduced by Wong *et al.* This method enables less trusted servers to compare the distances between different points over encrypted texts, i.e., the cipher texts, which was partially used to develop the privacy-preserved biometric identification method.

A person's face is a very complicated and dynamic object and its biometric elements are widely used in the recognition of individuals. Recently, biometric systems have been recognized in terms of security since the recognition of users and matching with listed ones are performed daily. To facilitate this process, the organizations must support them by using great computing tools. Some studies use the Principal Component Analysis (PCA) method for face recognition using cloud computing Maryam M Najafabadi [10].

In Chen and Zhao, a model of security of data is analyzed. A huge group of people are searching for data security on cloud platforms because the trust lessens because of the nature of the technology and it is the most important thing to gain the trust and confidence of users P. Failla *et al.* [13].

Phillips *et al.* [14] have reported that, traditionally, biometric identification methods have a number of advantages over different methods.

Xue-Wen *et al.* mainly targeted two factors. The primary focus is to determine the way that deep learning is able assist in solving issues in big data analytics specifically. The second target is to find out how to improve particular issues of Deep Learning to reduce a few of the problems linked with big data analytics. In this paper, the authors give a small overview of Deep learning and also showcase the recent research efforts and problems in big data, as well as future trends. The issues and problems related to the mining of big data are presented. This paper also discusses big data analysis tools like Hadoop Map Reduce and HDFS. Some security and privacy challenges related to big data are also described.

Ratha *et al.* describe the "four V" (Volume, Variety, Velocity, and Veracity) challenges of biometric big data and the representative techniques addressing these challenges using different diverse and realistic biometrics applications. Indexing methods are used for dealing with velocity and volume of the biometric database. The authors in their paper describe two indexing methods: one for fingerprints and the other for irises. Risk and accuracy concerns related to biometrics identification for large databases are addressed by using multiple (variety) biometrics. Four categories of Deep Learning (DL) architectures, such as auto encoder models, Restricted Boltzmann Machines models, Convolutional Neural Networks, and Recurrent Neural Networks are proposed. Some research trends and potential are also discussed.

11.3   Biometric Identification in Cloud Computing

Cloud Computing is one of the newest technologies. The user can the files or data on clouds from anywhere all over the Internet. The use of Cloud Computing has its own advantages, such as high availability, low costs, and so on. On the other hand, there are some security shortcomings in this field. One of them is the cloud usage time recognition of the authorized user; authentication must be provided for the user to access the clouds. Authentication is the process of verification and identification of the user by previously provided information [9].

In biometric recognition systems, the database owner who manages the fingerprints database sometimes wants to outsource the huge biometric dataset over to the cloud server (e.g., Amazon) to get away from expensive computation and storage costs. However, to ensure the security of biometric data, the data should be in the cipher text format and secured before sourcing it out. Then, whenever any partner or any other entity (e.g., the police station) wants to verify or match any user's identity, he asks the owner and generates an identification query only by using the individual's biometric traits (e.g., facial patterns, irises, voice patterns, fingerprints, etc.).

The database owner processes the encryption of the query before processing and then submits it to cloud server in order to find the appropriate match. The problem in this is how to make and develop a protocol or methodology which can enable efficient, easy, and privacy-preserving biometric identification for cloud computing [10].

Various privacy-preserving methods which are secure in biometric identification have also been introduced, but most of the methods focus on the preservation of the privacy of the dataset.

One of the main reasons why biometric authentication is so successful is it uses human traits that are unique to each individual. Traits like fingerprints and face recognition are used in biometric authentication and there is no other way one can access the cloud platform. Although both face recognition and fingerprints are used in biometric solutions, the latter one is more widely used. You should know that the concept of biometric identification first started with fingerprints and later on, with the advancement of technology, face recognition also become a part of biometric authentication. The ridges and valleys of fingerprints are unique to each individual

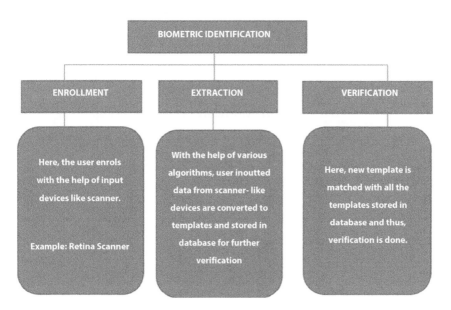

Figure 11.4 Biometric identification flow chart.

and that is why it was the first mode of authentication used in biometric authentication shown in Figure 11.4 [11].

11.3.1 How Biometric Authentication is Being Used on the Cloud Platform

When it comes to the cloud platform, then the cloud provider enrolls the cloud user in the biometric system for biometric authentication. During the process of authentication, multiple biometric fingerprints are enrolled and then saved in the form of templates in the cloud provider's section. In order to access the cloud application, the user is required to provide his fingerprint which is matched with the templates stored by the cloud provider in the biometric system. If the fingerprint provided by the user matches with the templates stored by the cloud provider, then the user is granted access [12].

This is how the biometric system comes into action whenever a user accesses his cloud application. Once the authentication is successful and completed, then there is a redirection where the individual is now sent to the actual cloud service platform. It is seen that users working on a cloud platform with a biometric system have a lesser chance of becoming a victim of any type of security breach in comparison to those who use traditional authentication processes shown in Figure 11.5.

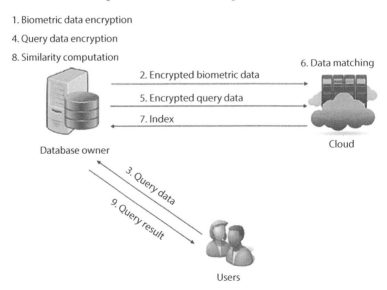

Figure 11.5 Biometric architecture of the cloud-based biometric identification system.

11.4   Models and Design Goals

11.4.1   Models

11.4.1.1   System Model

Just consider a biometric identification system based on cloud and consisting of three entities involved: Client, Cloud Server, and a Database Owner. The Database Owner haS the gathering of biometric data (D) whose copy he stores on a cloud server in an encrypted text format (C). For making data searches easier and faster, the Database Owner uses indexing methodology and creates index (I) for all the items in C and D, where a data item will have its one index which will be unique. The candidate's biometric data will be with the Client who wants to check whether the pattern data can be matched by database D or not. To match identification, the Client will send his biometric pattern data to the Database Owner [13]. The Database Owner then creates an authentication detail or data for the biometric data received and submits the credential to the cloud. The cloud servers will search for the appropriate match to the user's biometric data with the given authentication details and then returns the corresponding index of the data to the Database Owner. Thereafter, the plain text can be accessed by the Database Owner to make a match of the item using its index with the received data item and matches it with the previous model defined level, then sends the final authentication response back to the Client. Thus, we can see that, in this system model, time consumption of the data matching which is sourced out to the cloud is reduced.

11.4.1.2   Threat Model

Cloud servers are very efficient and fast, but at the same time are prone to many security threats due to their 24 hours network availability. The cloud platform protocols are well defined for better user efficiency and storage methods with less consumption cost [14], but there are ncreased threats as well which need to be addressed. The encrypted search on the cloud is one such issue. Let us assume that a threat or the attacker has access to the database of the cloud which is encrypted as well. Now, he will have the details of the user's biometric data and all the results which are intermediately created by the cloud servers. Now, for the better evaluation of the cloud servers and their strength, let us follow and extend the model of attack which classifies the three levels of the attackers:

Level 1: Here, the attacker finds only the credential of the candidate's biometric data and encrypted database C

Level 2: Besides the encrypted data, the authentication data of the database D, still does not have the respective encrypted keys of those rows in C

Level 3: Besides the details of the user's biometric data and encrypted biometric database, the attacker finds a set of rows in the database D and gathers the respective encrypted data of such rows in C

In these levels of the attack, the one at the higher level is a very strong level of attack made and is very good in terms of the data security. If the method at last level can be avoided, then it automatically shows security of the data at previous levels. Apart from Wong *et al.,* our model lets many users themselves become attackers and they can even collude with cloud servers to ensure the privacy of the data. This can be extended in a Level 3 attack by letting the attacker select plain text, so this interest in encryption, which goes according to the chosen message in the model of attack, is provided in the cryptography.

11.4.2   Design Goals

In order to enable privacy-preserving and required biometric identification on the cloud platform, our cloud service should fulfill the required security and performance requirements simultaneously:

- Cloud severs should be allowed by the system to work and perform most of the identification operations and the computational and communication cost should be kept at a minimum for the database owner as well as for the users using it
- The biometric data integrity and security should also be protected during this whole identification process in such a way the others private biometric data cannot be derived in any form [15]
- The system has to be efficient as well as scalable. In particular, the communication cost and the identification time for each candidate's biometric data should be independent to the size of database. A large number of request handling mechanisms should also be efficient in order to use serve all the requests.

11.5 Face Recognition Method as a Biometric Authentication

In modern times, the use of computer search systems for human face recognition is widely applied. In the network environment, different methods are used for human recognition, which results in improved recognition. This depends particularly on the capacity of the database. Note that when various databases are integrated in the network and the number of patterns stored in them is high, the recognition process becomes even.

The biometric features of human speech and facial expressions can be actively used in solving operative-search issues. Traditionally, biometric identification methods have a number of advantages over different methods:

- biometric signs are very difficult to be falsified
- identification reliability is very high due to the uniqueness of biometric signs
- biometric identifiers cannot be forgotten as a password or lost as a plastic card.

Safety is very difficult to be measured quantitatively. The increasing threat of terrorism and the need to improve security systems have led to the recent increase in volume and more challenges in the biometric equipment market.

The principal customers of biometric systems include not only commercial enterprises, but also government agencies. Particular attention is given to the places where mass surveillance systems are needed, such as airports, stadiums, and other facilities. Elements of the face of any person are individual biometric characteristics. The photos attached to documents are the most widely used for identification and related standards are adopted for all countries. Biometric identification based on photographs has many advantages as follows:

- The photo or video of a person's face does not require direct contact with the person and can be considered confidential in the course of the identification
- Databases of photographs are more accomplished and widespread rather than other biometric databases
- The application of this technology is not related to serious changes to state legislation
- Uniqueness and relative stability of inherited biological parameters of a human.

A person's face is a very complicated and dynamic object, as its features are used as biometric elements in the recognition of individuals. Recently, biometric systems have been recognized in terms of security, since recognition of users and their matching with listed data is performed daily. To facilitate this process, organizations must support them by using great computing tools. Some studies use the Principal Component Analysis (PCA) method of face recognition using Cloud Computing that it is implemented with the use of Scaling-Invariant Feature Transform (SIFT) by reducing the number of invariant points.

Face recognition is one of the most important aspects of ensuring security. Currently, experts mainly focus on monitoring in this area. Some articles suggest monitoring for face recognition in documents to be performed in cloud systems [16].

It consists of a customer module of monitoring for a Cloud Computing module for face identification and a double template is proposed. This method supports substantial local functions, reduces the difficulty for the Gabor filter, provides a periodical stability on texture and more accurate information detection and data visualization. In this case, Gabor's localization and segmentation of the centralized signs into a double template is proposed.

Digital data is growing exponentially due to advancements in web, mobile, and sensing devices. As per the National Security Agency, processing by the internet per day is 1,826 Petabytes. Big data is a term used to identify large size datasets with great complexity. Advanced analytics and visualization techniques are used on the huge data groups to find the hidden patterns in them.

11.6　Deep Learning Techniques for Big Data in Biometrics

Biometrics systems are big data systems due to the large volume of data and analytics involved in building real-world biometrics. Such systems handle applications which make use of citizen identity cards, voter IDs, and social benefits, as well as secure online and mobile payments. Biometrics are the authority center to recognize management and are going to be the hub of the next trend of cognitive systems.

Deep Learning techniques are currently the most popularly used for identifying objects in images and words in sounds. The successful application of Deep Learning in pattern recognition motivated researchers to use these techniques in difficult situations like translation of language automatically, medical diagnosis, and various other important issues.

There are hundreds of millions software governmental applications. Biometrics systems need to deal with not only a large number of records, but also a variety of multimodal data types, such as text, 1D signals, images, and video.

In biometrics systems, there are challenges regarding how to manage the fastest dynamic dataset which has entries at almost every microsecond, the fastest access to the required information, various security mechanisms of the system, and the confidentiality and integrity of the data records.

The forms of trusted identity systems, both 1:1 and 1:N, are normal sets of big data systems. Big data sets are linked with four Vs, which are variety, volume, velocity, and veracity. These dimensions are also associated with biometric data. In modern biometric systems, there is a lot of enrollment and verification data and the systems are designed in order to maintain such huge data operations.

Various software was developed using big data technologies for purposes such as criminal identification, face recognition, etc. Developers tried to implement a system that can search the users who were identified as thieves.

11.6.1   Issues and Challenges

The main four Vs characterized by big data systems are volume, variety, velocity, and veracity. These dimensions are also associated with biometric data as explained below:

1. Volume

The data involved in biometric mechanisms is huge and the dataset involved has tons of entries. The authentication and the identification of faces is a big issue that it has huge volumes to tackle and search from--millions and millions of data rows. Another major challenge is to resolve the 1:N mapping in searching and mapping requests rather than the 1:1 requests which are a better outcome.

Techniques to reduce dimensions aim at finding and exploring data structures with lesser dimensions in highly dimensioned data and save on computation and storage costs. The most popular and commonly used dimensionality reduction systems are Principal Component Analysis (PCA) and Linear Discriminant Analysis (LDA).

2. Variety

Biometrics systems are designed to run on a large variety of data types like image data, video data (faces or gaits), etc. Since the data coming from various sources are of different types, it is always a tedious and complex job to learn with such a heterogeneous data set.

To resolve the issue of uneven data integration, representation learning is preferred. In representation learning, data representation from each data source is learned first. The learned features at different levels are then integrated. Data from different heterogeneous sources can also be integrated very effectively using Deep Learning methods.

3. Velocity

Data processing speed is referred to as the velocity of the data in big data terminology. This velocity property of big data becomes very important in banking and ecommerce technology as their daily transactions and daily authentication entries reach the millions in numbers. This becomes another challenge when we need to make alerts and analysis from the same huge data set at the same time for a large number of users using the system at the same time. In a real time situation, the tasks have to be completed in the required amount of time within the limited memory constraints or else the results after a long time of processing become useless. In this scenario, Deep Learning comes into the picture and can be used to reduce the time frame of the operations.

4. Veracity

The veracity aspect largely worries about how to manage the basic failure occurrences or failure rates in the system, especially between false positives and false negatives [3]. Truthfulness can be maintained by using trusted enrollment, trusted verification, and identity credential management. In biometric systems, the biggest necessity is for the input signals to be coming live from the users in the real time.

The accuracy and trustworthiness of the initial level of data becomes a problem now a days because data comes from multiple heterogeneous sources and data quality is not all verifiable. Uncertainty and incompleteness are also associated with data quality. With incomplete data, correct interpretations and predictions of data are not possible, so to tackle these challenges, advanced Deep Learning methods can be applied.

11.6.2   Deep Learning Strategies For Biometric Identification

Machine Learning is the most emerging technology of the time. It actually aims at making machines learn about the respective tasks in a way a human would do in the same situation. This is achieved by feeding a large dataset of predefined situations to Machine Learning algorithms, making them learn in order to process whenever a new entry reaches it. Similarly, the same mechanism and data pattern is used for the predictions as well. At the same time, it also learns the new data pattern which is not fed before and makes the system advance. From the beginning, Machine Learning gave rise to neural networks and Artificial Intelligence. Their extreme use and the goal to provide the fastest computations led to the emergence of the term Deep Learning which is the inclusive of Machine Learning, Artificial Intelligence and neural networks [17].

Various types of Machine Learning are explained below:

(i)   Supervised Learning
(ii) Unsupervised Learning
(iii) Reinforcement Learning
(iv) Deep Learning

(i) Supervised Learning
The labelled dataset is the training set for supervised learning. Supervised learning finds an association with the set of features, labels it, and predicts the class label of test instances using training datasets. The data that comes out of supervised learning is generally used for predictions and recognition.

There are mainly two classes of supervised learning problems and they are regression and classification [18]. Different algorithms are available for solving regression and classification problems:

- Linear Regression
- Support Vector Machines

For big data analytics, efficient and advanced supervised methods suitable for parallel and distributed learning are required. Divide and conquer SVM, distributed decision trees, and neural networks are supervised learning algorithms suitable for big data, in which SVM is the most efficient and commonly used method shown in Figure 11.6 and Figure 11.7.

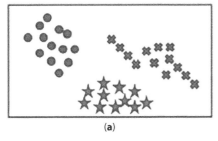
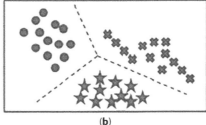

(a)   (b)

Figure 11.6 (a) Set of features of dataset values. (b) Prediction of class labels for the features.

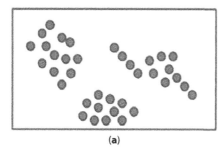 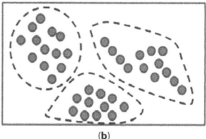

(a)   (b)

Figure 11.7 (a) Set of features having no predefined labels. (b) Cluster formation based on distance calculation.

(ii) Unsupervised Learning

The unlabeled dataset is used in unsupervised learning and the training dataset has no defined data labelling. The various clustering, such as kNN, hierarchical clustering, etc. comes under the category of unsupervised learning [19].

Unsupervised learning algorithms:

- K-means are commonly used for solving clustering problems
- Apriori algorithms are suitable for association rule learning problems

(iii) Reinforcement Learning

Reinforcement learning is a computational method to learning which tries to learn from feedback.

(iv) Deep Learning

Deep Learning the highest level in machine intelligence and is the most advanced technology, which needs high computation for faster data processing. It covers all Machine Learning, Artificial Intelligence, and neural networks. It does not depend on particular labeled or structured datasets. It uses unsupervised and supervised learning strategies. Big data processing is made and used in Deep Learning technology. The volume and variety characteristics of big data are focused on Deep Learning mechanisms, so Deep Learning is more suitable for processing high volumes of unstructured and heterogeneous data. Big data can be more accurately predicted using Deep Learning. Deep neural network models can be trained efficiently using scalable parallel algorithms.

The layout of a Deep Learning architecture consists of input layers, hidden layers, and output layers. Input data is given to the input layer and, in turn, is divided into multiple samples for data abstractions. The extraction of features from multiple levels and predictions is done by the intermediate

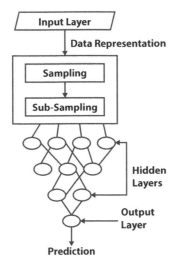

Figure 11.8 Deep learning architecture layout.

layers. The outputs from the intermediate layers are forwarded to the output layer for final prediction shown in Figure 11.8.

11.7 Conclusion

Biometric identification on the cloud has been analyzed by many studies and research papers about its confidentiality, integrity, and availability. In this paper, biometric authentication is coupled with many encryption techniques to assure the security of different biometric techniques like image and face reorganization on the cloud architecture. Biometric techniques are gaining popularity as authentication techniques to authorize users and fix many permission checks to stop other unauthorized entities from getting in. Biometric authentication techniques as discussed above are recommended for authorization if the process of authorization itself is strong and errorless for ensuring the security of accessing clouds in the future. This paper describes a brief survey on various approaches of Deep Learning for handling big data in biometrics is presented.

References

1. Abhishek Kumar & Jyotir Moy Chatterjee & Pramod Singh Rathore, 2020. "Smartphone Confrontational Applications and Security Issues,"

International Journal of Risk and Contingency Management (IJRCM), IGI Global, vol. 9(2), pages 1-18, April.

2. Barbara Hammer, Haibo He, and Thomas Martinetz, "Learning and modelling Big data", ESANN 2014 proceedings, European Symposium on Artificial Neural Networks, Computational Intelligence and Machine Learning. Bruges (Belgium), 23-25 April 2014, i6doc.com publ., ISBN 978-287419095-7.

3. Bhargava, N., Bhargava, R., Rathore, P. S., & Kumar, A. (2020). Texture Recognition Using Gabor Filter for Extracting Feature Vectors With the Regression Mining Algorithm. International Journal of Risk and Contingency Management (IJRCM), 9(3), 31-44. doi:10.4018/IJRCM.2020070103.

4. David Menotti, Giovani Chiachia, Allan Pinto, William Robson Schwartz, Helio Pedrini, Alexandre Xavier Falc˜ao, and Anderson Rocha, "Deep Representations for Iris, Face, and Fingerprint Spoofing Detection", IEEE.

5. George Chellin Chandran. J, Dr. Rajesh. R. S., "Performance Analysis of Multimodal Biometric System Authentication". IJCSNS International Journal of Computer Science and Network Security, Vol.9 No.3, March 2009.

6. J. de Mira, Jr., H. V. Neto, E. B. Neves, and F. K. Schneider, "Biometric-oriented iris identification based on mathematical.

7. Jaseena K U and Julie M David, " Issues, Challenges, and Solutions : Big Data Mining", NeTCoM, CSIT, 2014, pp. 131– 140.

8. Pawle and Pawar, "Face Recognition System (FRS) on Cloud Computing for User Authentication" International Journal of Soft Computing and Engineering (IJSCE) ISSN: 2231-2307, 3, 4, September 2013.

9. P Rathi, S Dubey. Int J Adv Res Comput Eng Technol 2 (6), 2064-2067, 2013. 9, 2013. Methods to ensure quality of service in cloud computing environment.

10. Maryam M Najafabadi, Flavio Villanustre, Taghi M Khoshgoftaar, Naeem Seliya, Randall Wald, and Edin Muharemagic, "Deep learning applications and challenges in big data analytics", Journal of Big Data, USA, 2015, vol. 2, No. 1, pp. 1-21.

11. N. Bhargava, S. Dayma, A. Kumar and P. Singh, "An approach for classification using simple CART algorithm in WEKA," 2017 11th International Conference on Intelligent Systems and Control (ISCO), Coimbatore, 2017, pp. 212-216, doi: 10.1109/ISCO.2017.7855983.

12. Naveen Kumar, Prakarti Triwedi, Pramod Singh Rathore, "An Adaptive Approach for image adaptive watermarking using Elliptical curve cryptography (ECC)", First International Conference on Information Technology and Knowledge Management pp. 89–92, ISSN 2300-5963 ACSIS, Vol. 14 DOI: 10.15439/2018KM19.

13. P. Failla, D. Fiore, R. Lazzeretti, V. Piuri, F. Scotti, and A. Piva, "Privacy-preserving fingercode authentication," in Proceedings of the 12th ACM

workshop on Multimedia and security, ser. MM&Sec'10. New York, NY, USA: ACM, 2010, pp. 231–240.

14. Phillips *et al.*, "An Introduction to Evaluating Biometric Systems, Guide to Biometrics", IEEE Computer, February 2000, pp. 56-63. K. Jain, A. Ross, and S. Pankanti, "A Prototype Hand GeometryBased Verification System", 2nd Intl Conference on Audio- and Video- based Biometric Person Authentication, Washington D.C., pp. 166-171.

15. R. Allen, P. Sankar, and S. Prabhakar, "Fingerprint identification technology," in Biometric Systems. London, U.K.: Springer, 2005, pp. 22–61.

16. R. Bolle and S. Pankanti, Biometrics, Personal Identification in Net- worked Society: Personal Identification in Networked Society, A. K. Jain, Ed. Norwell, MA, USA: Kluwer Academic Publishers, 1998. M. Barni, T. Bianchi, D. Catalano, M. Di Raimondo, R. Donida Labati.

17. Salil Prabhakar, "Fingerprint classification and matching with filterbank", Ph.D Thesis, University of Michigan State, 2001.

18. Shafagat Mahmudova, "Big Data Challenges in Biometric Technology", I.J. Education and Management Engineering, 2016, 5, 15-23 Published Online September 2016 in MECS http://www.mecs-press.net).

19. Shui Yu, "Big Privacy: Challenges and Opportunities of Privacy Study in the Age of Big Data", IEEE, Volume 4, June 2016.

20. Yassin, M. & Alazba, Prof & Mattar, Mohamed. (2016). Comparison between gene expression programming and traditional models for estimating evapotranspiration under hyper arid Conditions. Water Resources. 43. 412-427. 10.1134/S0097807816020172.

12

Application of Deep Learning in Cloud Security

Jaya Jain

MDS University, Ajmer, India

Abstract

As we have seen over the years, Deep Learning has become a pattern. An in-depth study helps improve research and can learn a large amount of unparalleled information. Additionally, data protection procedures will be limited to a few fields. As a result, this paper provides an in-depth study survey and an idea for a new application for in-depth study in future research. On the other hand, various methods data protection for Cloud Computing have been proposed. However, there are many drawbacks to this issue. The security point improves cloud security and data efficiency in cloud outsourcing, as security is an important issue in the cloud and a proper algorithm is needed. With the cloud, security is secure. This method is based on two algorithms, one based on machine learning and the other on a neural system. The machine learning algorithm is based on a KNN algorithm and neural system strategy. Based on data fragment and hashing technology, both of these processes should optimize cloud security by using cloud data encryption to the cloud server. There are seven applications used for the in-depth study described, namely, customer relationship management, image recognition, natural language processing, recommendation programs, automated speech recognition, drug discovery and toxicology, and bioinformatics. In this case, we will discuss the results of the study and identify areas that need to be redesigned.

Keywords: Advanced reading, applications, access control, authorization, cloud

Email: jayavineetjain@gmail.com

Pramod Singh Rathore, Vishal Dutt, Rashmi Agrawal, Satya Murthy Sasubilli, and Srinivasa Rao Swarna (eds.) *Deep Learning Approaches to Cloud Security*, (189–206) © 2022 Scrivener Publishing LLC

12.1   Introduction

In-depth study on cloud security:
Deep Reading (DL), also known as systematic reading or hierarchical reading, is a machine-readable (ML) division based on a set of algorithms that attempt to mimic high-output data. Such algorithms improve the optimal textual structure of reading and data representation. This method of constructing learning is inspired by artificial intelligence that mimics the deep, systematic learning process of important neocortex regions in the human cerebrum, which removes results and errors from the data below (3,4,5). Based on (6,7), DL algorithms are helpful in dealing with large amounts of unattended data and naturally read data presentations in a selfish and intelligent way. In recent years, there have been various researchers using the DL algorithm and definitions.

Definition 1
An in-depth study is a collection of algorithms in machine learning that attempt to learn at many steps, corresponding to different steps of extraction. It usually prefers artificial neural systems. The levels in these types of learning are aligned with different levels of conception where high-level concepts are defined from low-level meetings and concepts of the same high level can help define many high-level concepts [1].

Definition 2
In-depth learning is another form of machine learning, given the purpose of drawing Learning Machines close to one of its distinct purposes. In-depth learning involves learning different steps of presentation and dissipation that help to understand details, for example, pictures, sound, and text.

In recent years, there have been a number of researchers using DL techniques in a variety of fields and, therefore, the purpose of this paper is to review and discuss applications and cloud security made over the years with DL algorithms and cloud security.

Advances in technology have taken computing to a whole new level and one of the most recent developments in this context is the introduction of cloud computing. It has changed the concept of distributed computers and, thanks to advanced technology, high performance computers are cheaper and more available and Cloud Computing means the distribution of computers, including hardware and software, to the consumer online. Cloud Computing can be divided in two ways: by cloud computing location and by the types of services offered.

With cloud acquisition, Cloud Computing is often divided into the public cloud (where computer infrastructure is hosted by a cloud vendor), the private cloud (where computer infrastructure is provided by a particular organization and can be shared with other organizations), the hybrid cloud (the use of private and public clouds together), and the public cloud (including IT infrastructure sharing between organizations of the same community).

It is still distributed in a major segment, such as a platform as a service (PaaS), software as a service (SaaS), or infrastructure as a service (IaaS) [2]. Although the selected model is determined by what is delivered as a service, the main context remains the same for the distribution of computer resources by export.

12.2   Literature Review

Research has been done to create reasons for the analysis of huge data in the cloud. Data stored in deep reading can be thought of as large data as it includes numbers of sets of photos, videos, and audio. Appropriate papers have tried to clarify the use of the cloud platform in this regard.

The focus of the paper written by Zai, Chun-Wei, *et al.* [37] was a question on how to develop a high-performance platform for high data analytics. It also tries to design an algorithm that can get useful information from huge data.

The paper Salloumet *et al.* studies analytics on the Apache Spark platform. A comprehensive data analytics framework with its advanced form of in-memory memory, top-of-the-line machine learning libraries, graph analysis, streaming, and structured data editing [3].

Another paper written by Middelfart and Morten [27] performs cloud computing analysis and business intelligence using two different methods, i.e., a professional analysis and communication field.

Mohammad, Atif, *et al.* [28] came up with Big Data Architecture, considering its relationship to Analytics, Cloud Computing, and Business Intelligence.

Fiore, Sandro, *et al.* [16] have proposed a cloud-based data analysis infrastructure for biodiversity change.

Hamdaqa, Mohammad, *et al.* [17] proposed the Ad-hoc Cloud Map Computer Framework. Analysis of new data frameworks in the cloud is proposed. Big health care data analysis is the task of obtaining data on the largest datasets used to improve health services.

Farheen discussed Cloud Computing methods for huge data.

This large amount of data puts a lot of pressure on the performance of the script, the disability, the need for effective storage, and the rational operation of this data. Indigenous knowledge data is not enough for this. It was therefore designed to rebuild the big data structure that includes the application collection, storage component to enable reading, writing, and refreshing speed, and data growth.

The paper is designed for in-depth Convolutional Neural Networks (CNNs) to incorporate into the deeper aspects of the cloud [4].

Work has been done to predict a VM workload in the cloud using in-depth reading, establish an effective cloud system for in-depth learning, 3D point data extraction features using a default installer, and design deep depth models to reduce costly operations in the cloud.

Some studies show the benefits of hardware acceleration and higher gains in the cloud.

12.3   Deep Learning

In-Depth Learning Approach

A. Auto-Encoders

Many of the advanced deep constructions include knowing the RBM layers or Auto-Encoders for each of these two internal neural networks that find ways to convert inputs [5].

The RBM type for their input is a distribution of opportunities, while Auto-Encoders find ways to reproduce. RBM is an unbalanced neural system similar to the visible layer and hidden layer. There is no connection between each layer, however, the connection appears to be hidden and trained to maximize expected data entry. Inputs are binary packages, as read by Bernoulli's distribution across all entries [6].

The activation function is calculated in the same way as a normal neural network and the standard order characteristic commonly used is between 0-1.

Outcomes are treated as an opportunity and all neurons function when the activity is larger than random variance. Hidden neurons take on visible units as input [7].

Visual neurons take several input signals as the first input and the potential for potential encryption. Most more advanced modern engineering involves knowing the RBM layers or Auto-Encoders for each of these two internal neural systems that detect input modes.

An RBM version of their input is a shared opportunity, while Auto-Encoders find input and output modes, such as output.

In the training phase, Gibbs Sampling (MCMC process) has been completed and compared to the standard distribution of the Markov Chain Monte Carlo Method shown in Figure 12.1 [8].

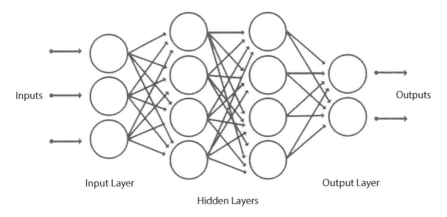

Inputs

Outputs

Input Layer

Output Layer

Hidden Layers

Figure 12.1 Deep learning strategies.

With step 1, the hidden layer of probability (h) is calculated from the input (v) and with step 2, those numbers go back to the visible layer and return the hidden layer to find the v and h.

Weights change using the difference within external products for hidden and visible performance between the first and second passes. Approaching the right model, a large number of passes are required, so this method provides subtlety, but works good in practice [9]. The hidden RBM performance can be used as learning features after the training.

The automated encoder is a neural feed supply community that aims to explore a dynamic, scattered image of data. The automotive encoder is a three-dimensional neural system, with the ability to rearrange its input, using it as a result.

If line-based operations are better and can also be used to decrease the size, they can present as similar to PCA [10]. Once trained, the use of the hidden layer is used because the learned function and upper layer can be removed.

Automatically driven encoders are trained in the use of techniques such as denoising, contraction, and sparseness. In Auto-Encoders some random sound is transmitted to the input. The encoder is required to reproduce the original input during denoising.

Random input will do better than the normal functioning of the normal neural system in training.

On automated computers, setting the width of the enclosures within the hidden layer is much lower than the number of input nodes forcing the public to make a decrease in size. This prevents access to the identification function because storage space is not sufficient within the hidden layer [11].

Sparse Auto-Encoders are trained in the use of the sparsity penalty in weight update work. Slow down the total length of the connecting weights and many heavy reasons to have smaller values [12].

RBMs or Automatically-driven Encoders can be layer-based. The techniques learned in one layer are transferred to the next sections, so that the network with the hidden layer is in step one trained and, only after this is achieved, a network with two hidden layers is trained, and so on. In each step, a network of vintages with one hidden layer is taken and given a hidden k-th layer which acts as a pre-hidden pre-1 layer to be trained.

B. Convolution Neural Networks (CNN)

Convolution Neural Networks (CNN) are an MLP-inspired variety [13]. A standard Convolution Neural Network is integrated for various layer areas, other feature presentation layers, and others as a common type of neural segregation network. Convolution and subsampling layers are the two types of layers.

These layers perform binding operations with a combination of multiple filter maps of the same size and neighboring size, while layers of smaller samples decrease the size of continuous layers by medium pixels in a small area shown in Figure 12.2.

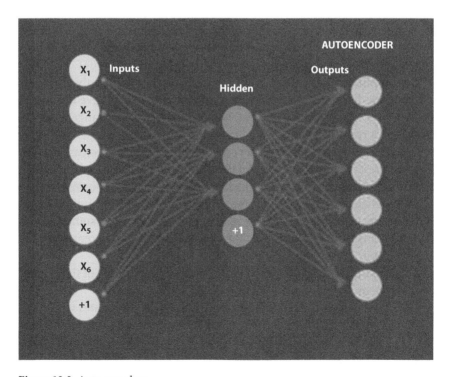

Figure 12.2 Auto-encoders.

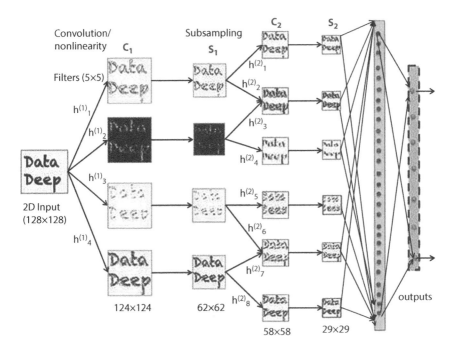

Figure 12.3 Data deep feature maps.

Input is first bound by a sorted order. These refined 2D records are called feature maps. After a nonlinear transformation, the sample below is added again to reduce the size and the confirmation of the acquisition or the sample below can be repeated repeatedly. The lower the level of this formation, the lower the insertion layer. With local receiving fields, the upper layer of neurons is one of the most complex objects. The whole identification layer is made without the few commercial signals that are formed using the input (S) with different filters. In other words, the importance of the whole unit on the feature map is the result of depending on the internal reception area within the previous frame and filter. The sequence of character themes is converted into a word shown in Figure 12.3.

CNN algorithms learn management representation through strategies such as local acceptance platforms, shared weights, and subsamples.

12.4   The Uses of Fields in Deep Learning

This section explains the basis for this review by discussing the several fields it has intensively used with the algorithm.

A) Speech Recognition
B) Bioinformatics
C) Customer Relationship Management
D) Drug Disposal and Poisoning
E) Image Recognition
F) Performance of Indigenous Languages
G) Recommendation Programs

A) Speech Recognition (SR)

Google announced that Google Voice Search has taken a new step in adopting Deep Neural Networks (DNN) as a technology used for the sound of voice sounds in 2012 [8]. DNN has replaced the Gaussian Mixed Model that has been in the industry for 30 years. DNN has also proven to be able to better measure which user is always creating at a time and, with this, they have presented visual accuracy of the speech. In 2013, DL gained full momentum in both ASR and ML [14]. DL is basically linked to the use of multiple layers of non-linear translation to acquire speech features, while learning about shallow layers involving the use of example-based presentation of high-quality speech, but not allowing input features shown in Figure 12.4.

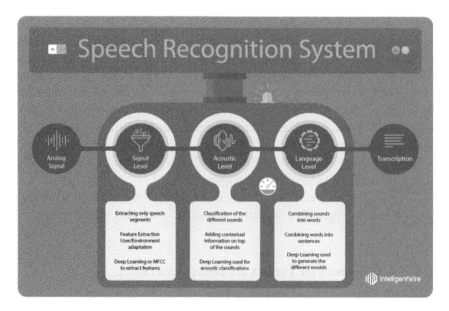

Figure 12.4 Speech recognition system.

- We can identify the following key steps:

  - Analog-to-Digital Conversion: Speech is usually recorded and available in analog format. Typical sampling techniques and devices are available to convert analog speech into digital sample size and quantity techniques. Digital speech is usually a single measure of speech signals, each of which is a whole number.
  - Digital speech is usually a speaker equal to the size of speech samples, each of which is a number.

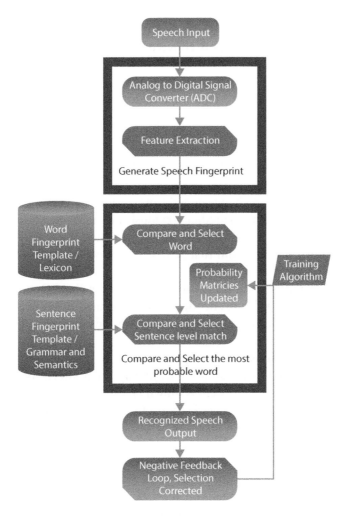

Figure 12.5 Feature extraction.

- Preprocessing speech: Recorded speech usually comes with background sound and a long sequence of silence. Prespeech adjustment includes identifying and removing silent frames and techniques for using samples to reduce and eliminate noise. After preliminary consideration, the speech was broken into 20ms frames each in order to take further action.
- Dispatch Feature: the process of transforming the speech impediment into a computer feature that determines which grammar and language system is being spoken [15].

Word choice: Based on the language model and sequence model, the sequence of character themes is converted into a word shown in Figure 12.5.

B) Bioinformatics

The definition of genomic data is a major challenge in biology and bioinformatics.

The existing data for known genetic functions is incomplete and prone to errors and the bimolecular tests required to improve this information are slow and expensive [16]. While computational methods are not a place of experimental validation.

Algorithms can help change genetic annotations by automatically providing inaccuracies and can predict previously unknown genetic functions, speeding up the detection rate of the genetic function.

In this work, an algorithm that achieves both objectives through deep auto encoder networks is developed.

Examination of genetic annotation data from the Gene Ontology project shows that standard auto encoder networks achieve better performance than other conventional machine learning methods, including the design of a single value preference.

C) Customer Relationship Management

First, the structure of the autonomous management system of customer relationship management system is drawn [17].

Second, how to get direct marketing action in an open and continuous area of individual action, in terms of its position in the public domain is explained.

Marketing Activities are the estimated prices above the customer space can be defined as the life expectancy of the customer (CLV). Introducing test results based on Competition for Competitiveness Tools and Data Tools, an email mailing page for donations.

D) Drug Discovery and Poisoning

Qualification analysis and prediction studies (QSAR/QSPR) attempt to construct mathematical models related to the physical and chemical properties of the elements that are built into their chemicals. In [18], multidisciplinary learning is included in QSAR using a variety of different neural models. They have used a neural implant network to learn a task that predicts computer typing tasks simultaneously.

This approach is comparable to other methods and it is reported that high-performance neural networks can lead to significantly improved outcomes over bases formed by random forests.

This paper also shows how you can apply the concepts of environmental elements and hierarchical structures to the mode of bioactivity and chemical interactions. Atom Net outperforms previous docking methods for various sets of benches with large margins, reaching an AUC of more than 0.9 to 57.8% of the downgraded stones on the DUDE bench.

In various terms, therapeutic activity occurs frequently. Diagnosis, which often occurs in one of the remedies, has seen an oceanic approach from the finger doctor to your local doctor using sophisticated information processing equipment obtained from biomarkers found in the form of slow attacks. One hundred years ago or so it proved to be a case of universal access to allopathic medicine, an act aimed at gaining preexisting preclinical practice and homeopathy.

In addition to these days, the new renewal that creates a normal life goes into health care. The discovery of drugs and the solution that started with Ayurveda in India is currently undergoing a special artificial intelligence (AI) strategy, which includes the discovery of new chemical entities and the excavation of data on highways protected by health. Indeed, various disciplines, such as language, neurophysiology, chemistry, toxicology, biostatistics, medicine, and computers, have come together to integrate algorithms primarily based on the transmission of information in neural

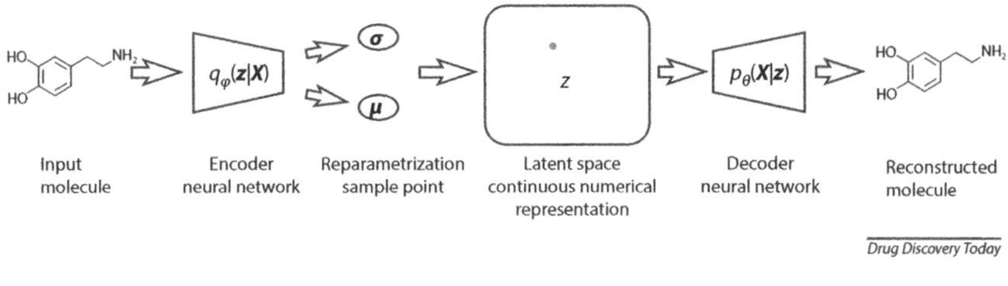

| Input molecule | Encoder neural network | Reparametrization sample point | Latent space continuous numerical representation | Decoder neural network | Reconstructed molecule |

Drug Discovery Today

Figure 12.6 Drug discovery.

networks that replicate the application of drug applications using novel, prior information of its protection, and approving and engineering their deployment is much more efficient than necessary, with all such listing and testing comic signatures on all diseases listed in the stages of history to allow basketball treatment strategies shown in Figure 12.6.

This review highlights the recurrence of drug-induced drug and cell therapy and in the field of integrated medicine system with molecular therapy. Cell therapy is used for advanced development, greatly assisted by inputs in comic analysis, cognitive medicine, big data capture and analysis, in-depth learning, and AI.

In fact, the wedding of AI-assisted drug discovery and appointment and drug-assisted drug use and patient management will change the field of medicine.

E) Image Recognition
Next, networks of max-pooling convulsive neural networks are used to detect mitosis in breast histology images [10]. Diagnosis of mitosis is very difficult. In fact, mitosis is a complex process in which the nucleus of a cell undergoes many changes. In this way, DNN as a powerful pixel classifier works with raw pixel values and no human input is required. Therefore, DNN automatically reads a collection of visual elements from

Figure 12.7 Image recognition.

the training data. DNN is tested on publicly available databases and is very effective in all competitive strategies with controllable management effort; processing a 4MPixel image requires a few minutes on a standard laptop shown in Figure 12.7.

A large and deep network of neural faith is trained to distinguish a high resolution of 1.2 million images in the Image Net LSVRC-2010 competition in 1000 different categories [19].

For test data, they found 1 and top-5 errors of 37.5% and 17.0%, which is much better than the previous state. In all tests, results can be improved by simply waiting for faster GPUs and more comprehensive data to be available.

F) Performance of Indigenous Languages

Recently, an in-depth learning process has been used successfully in a variety of language and information applications. By exploiting deep construction, in-depth learning strategies are able to derive from learning data the hidden structures and features at different levels of pull that are useful for any task. In 2013, [20] proposed a series of Deep Structured Semantic Models (DSSM) web search. Specifically, they use DNN to measure the text of a given query.

First, non- linear projection is made to look at a question with documents in the normal semantic area. After that, the value of each letter given in question is calculated as a cosine similarity between their veins in that semantic space.

Neural network models are trained in using click-through statistics by the fact that the chances of the text-based text given in question is increased.

Subsequently, new models were analyzed in Web document design work using a real-world data set.

The results show that the proposed model is significantly superior to other previous semantic models, which were considered to be up-to-date in performance before the work was presented [21].

What is NLP?

The functioning of natural languages (NLP) provides the building of critical analysis combinations that often analyze and represent human language. NLP-based systems fully enable a range of packages that include Google's powerful search engine and, more recently, an Amazon voice assistant named Alexa. NLP also helps teach machines the ability to perform complex language-bound obligations including gadget translation and speech time.

This reason is of concern, for example, when the magnitude when considering the linguistic details being translated is represented by sparse presentations (great skills). However, with the recent reputation and achievements of embedding phrases (small size, distributed presentations), level-based fashions have produced improved results in language-related activities compared to a standard fashion information device such as SVM or signal management [22].

G) Recommendation Programs

Automated music compliments have become a relevant issue in the current book because loads of songs are now being distributed and digitally fed. Many of the literature delivery facilities rely on collaborative filtering. In 2013, [23] proposes to use the latest recommendation model and predicts the latest features from the soundtrack after no data was available for use.

The traditional method is compared to using a voice wallet to broadcast radio signals with a deep neural system and the prediction is tested using the quantity and accuracy of a million song dataset.

The result reveals that recent advances in DL translate entirely into a system of musical compliments, with a deeper system of neural authentication needed than the traditional process.

The latest Internet connectivity services rely heavily on personalization to adapt to the active content of a large number of customers.

This requires an immediate rating system to address the movement of new users visiting online services. For the first time, the work performed by [24] in 2015 was proposed based on the content of the recommendation system for the quality of recommendations and the disability of the system.

They also propose using a rich feature to set up user representatives based on their web browsing history and questions they receive.

Scalability analysis shows that a multi-view DNN model can easily measure installation [25, 26].

12.5   Conclusion

As used in the conclusion of the section, there are seven applications for in-depth study identified. The applications are:

    a. Automatic Speech Recognition
    b. Bioinformatics
    c. Customer Relationship Management
    d. Drug Disposal and Poisoning
    e. Image Recognition
    f. Performance of Indigenous Languages
    g. Recommendation Programs

In addition, the chapter has discussed a different definition related to in-depth reading. In the next section, we described the literature survey. After that, we described the great reading technology. In this section, we have explained how to incorporate the output and output function, as well as the encoder and CNN and how both methods are useful for in-depth learning.

In the next section, we have discussed all seven tasks which are the most important in this chapter.

Deep Learning is a quick and uplifting discipline. It is difficult to build a good balance and a brief overview of the latest developments in the industry. Therefore, it is still very difficult to have a plan for future use.

References

1. Abhishek Kumar & Jyotir Moy Chatterjee & Pramod Singh Rathore, 2020. "Smartphone Confrontational Applications and Security Issues," International Journal of Risk and Contingency Management (IJRCM), IGI Global, vol. 9(2), pages 1-18, April.
2. Arel, Mina, Rose, D. C., & Karnowski, T. P. (2010). In-depth machine learning — a new frontier in between artificial intelligence study [research limit]. IEEE Intelligence Magazine, 5 (4),13-18.
3. Bengio, Y. (2009). Learns the deep structures of AI. Foundations and trends® in Machine learning, 2 (1), 1-127.
4. Bengio, Y., & LeCun, Y. (2007). Evaluating learning skills in AI. Large scale kernel equipment, 34 (5).
5. Bengio, Y., Courville, A., & Vincent, P. (2013). Read presentations: New reviews ideas. Release of IEEE in pattern analysis and machine understanding, 35 (8), 1798-1828.
6. Bengio, Y., Lamblin, P., Popovici, D., & Larochelle, H. (2007). Selfish training for selfishness deep networks. Progress in neural information repair programs, 19, 153.
7. Bhargava, N., Bhargava, R., Rathore, P. S., & Kumar, A. (2020). Texture Recognition Using Gabor Filter for Extracting Feature Vectors With the Regression Mining Algorithm. International Journal of Risk and Contingency Management (IJRCM), 9(3), 31-44. doi:10.4018/IJRCM.2020070103.
8. Chaudhuri, Surajit. "What's next?? Twelve data management targets for big data and cloud." Continuation of the talk of the 31st ACM SIGMOD-SIGACT-SIGAI series on Data Processing Procedures. ACM, 2012.
9. Chicco, D., Sadowski, P., & Baldi, P. (2014, September). Deep neural networks of autoencoder for gene ontology annotation annotations. In progress of the 5th ACM Conference on Bioinformatics, Computational Biology, and Health Informatics (pp. 533-540). ACM.

10. Ciresan, D. C., Giusti, A., Gambardella, L. M., & Schmidhuber, J. (2013, September). Mitosis detection in history of breast cancer with deep neural networks. Worldwide Conference on Photographic Therapeutic Science and Computer Intervention (pp. 411-418). Springer Berlin Heidelberg.
11. Dahl, G. E., Jaitly, N., & Salakhutdinov, R. (2014). Neural networks for multiple QSAR activities forecasts. arXiv prerint arXiv: 1406.1231.
12. Deng, L., & Li, X. (2013). Speech recognition learning parameters: An view all. IEEE Release in Audio, Speech, and Language Processing, 21 (5), 1060-1089.
13. Deng, Li, Geoffrey Hinton, and Brian Kingsbury. "New types of deep neural network learning for speech recognition and related applications: Overview." Acoustics, Speech and Signal Processing (ICASSP), 2013 IEEE International Convention on. IEEE, 2013.
14. Elkahky, A. M., Song, Y., & He, X. (2015, May). The standard reading mode for user domain model to opt out of recommendation programs. Following the 24th International Conference on the World Wide Web (pp. 278-288). ACM.
15. Farheen Siddiqui "Applications for Intelligent Distributed Processing Data" IOSR Journal of Computer Engineering (IOSR-JCE) ISSN: 2278-0661, p-ISSN: 2278-8727, Volume 18, Complaint 6, Ver . Mina (Nov. - Dec. 2016), PP 61-64 DOI: 10.9790 / 0661-1806016164.
16. Fiore, Sandro, et al. "A major data analysis of climate change and infrastructure variability in the EUBrazilCC cloud infrastructure." Continuation of the 12th ACM World Summit with Computing Frontiers. ACM, 2015.
17. Hamdaqa, Mohammad, et al. "Adoop: MapDovers ad-hoc cloud computing." Continuation of the 25th annual international conference on Computer Science and Software Engineering. IBM Corp., 2015.
18. Hasim Sak, Andrew Senior, Kanishka Rao, Françoise Beaufays and Johan Schalkwyk (September 2015): Google voice search: faster and more accurate.
19. Hinton, G. E., Osindero, S., & Teh, Y. W. (2006). A fast learning algorithm for deep belief nets. Neural Census, 18 (7), 1527-1554.
20. Hinton, G. E., Osindero, S., and T Te, Y. W. (2006). A fast learning algorithm for deep belief nets. Neural Census, 18 (7), 1527-1554.
21. Huang, P. S., He, X., Gao, J., Deng, L., Acero, A., & Heck, L. (2013, October). Warming in-depth structured web search models using click-through data. Following the 22nd day ACM international conference on information and information management conference (p. 2333- 2338). ACM.
22. Ian Goodfellow, Joshua Bengio, Aaron Courville, "In-depth Reading", MIT Press, 2016.
23. Krizhevsky, A., Sutskever, Mina, and Hinton, G. E. (2012). Imagenet segmentation in depth neural authentication networks. Advances in neural information processing systems (p. 1097-1105).
24. Kumar Sahwal, ..., Kishore, ..., Singh Rathore, ..., & Moy Chatterjee, ... (2018). An Advance Approach of Looping Technique for Image Encryption

Using in Commuted Concept of ECC. International Journal Of Recent Advances In Signal & Image Processing, 2(1).

25. Kumar, A., Chatterjee, J. M., & Díaz, V. G. (2020). A novel hybrid approach of svm combined with nlp and probabilistic neural network for email phishing. International Journal of Electrical and Computer Engineering, 10(1), 486.

26. Manish Kumar Pandey, Karthikeyan Subbiah, "The Structure of a Novel Preservation Plan for Healthcare Co-operation Advanced Information in the Cloud", INEEE International Convention on Computer and Information Technology (CIT) Year: 2016 pages: 578 - 585, DOI: 10.1109 / CIT.2016.86.

13

Real Time Cloud Based Intrusion Detection

Ekta Bafna

MDS University, Ajmer, India

Abstract

With the rapid development in the information technology, network technology, computer security, and cyber security, we need intrusion detection technology to defend from attacks. Despite that fact, numerous new advancements and security upgrades are adhered to the technology, penetrating the security measures to achieve the same development and improvements. To improve the accuracy of network intrusion detection and reduce false alarm, a Deep Learning algorithms defined from a KDD Cup 99's dataset is designed. This chapter explains the classification and prediction of network assaults through an algorithm. The real time intrusion detection system finds the network attack using the different Deep Learning algorithms to calculate the accuracy detection rate and false alarm rate and generate higher accuracy and efficient prediction of network attacks.

Keywords: Incursion in cloud, types of IDS, model of deep learning, KDD dataset

13.1 Introduction

Digital security incorporates a lot of frameworks and procedures to ensure network and computer system data and programs from unauthorized access and interference with organized attacks. Network security incorporates an arrangement of firewalls, antivirus programming, and an intrusion detection system (IDS). Interruption location is to gather and dissect information about the key nodes in a system to discover if there are abused security

Email: ekta.bafna026@gmail.com

Pramod Singh Rathore, Vishal Dutt, Rashmi Agrawal, Satya Murthy Sasubilli, and Srinivasa Rao Swarna (eds.) Deep Learning Approaches to Cloud Security, (207–224) © 2022 Scrivener Publishing LLC

practices or indications of being attacked. The Intrusion Detection System (IDS) is an autonomous framework giving the neighborhood arrangement a local network to guarantee the security of the system framework. There are lots of researchers applying AI strategies in IDS to recognize interruptions in both the scholarly community and industry. Many organisations are doing work for better security and are also consolidating the ongoing assaults, while new assaults are rising day by day [1].

In this lexicon, there are growing requirements for new explanation for predicting and preventing network interruption assaults progressively in the system.

The Intrusion Detection System is categorised into 3 parts: an anomaly based detection system, signature based detection system, and hybrid based detection system, which is a combination of anomaly intrusion and signature based intrusion. This detection framework depends upon the knowledge of arranging information on boundaries and the body content once an assault has happened. This irregularity detection is based on a recognizable proof structure to separate the assault payload data and other data, for instance, the source, objective port, and the web convention to develop the anomaly model [2]. This model will be produced when similar attacks come back into the framework. Similarly, these models are assembled depending upon past assault guidelines. Be that as it may, new attacks emerge in an increasingly compelling self-learning model and ought to be executed so as to empower the system to predict and prevent from concealed assaults.

The network intrusion detection framework has various strategies to anticipate intruder attacks over the web. NIDS uses a pattern to predict the attack type. NIDS datasets use the information of real time attack log file data to detect the type of attack. Network error handling is essential since it might cost the association (organizations). A Deep Learning model can be utilized by an intrusion detection system to classify and predict the network attack type on the basis of a network log-file dataset.

Cloud Computing has popular technology among the clients because of its huge advantages and applications in different areas. It permits the clients to configure, deploy, and manage services over the globe without the requirement for contributing to IT framework. Other than a few favorable circumstances presented by the cloud model, clients are fearful to give delicate data to the cloud because of the threat presented by hackers. Hence, the foremost responsibility of the service provider of the cloud is to provide maximum security to data and maintain the privacy of client information. The security structure has utilized strategies, similar

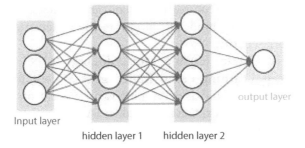

Figure 13.1 Deep learning model [13].

to firewalls, an intrusion prevention system (IPS), and intrusion detection system (IDS) to beat the security issues prevailing in the cloud shown in Figure 13.1 [3].

This chapter focuses on real time network intrusion detection by utilizing a Deep Learning algorithm. A cloud based model framework was created to examine the ability of a Deep Learning AI algorithm based network intrusion in a real time environment.

Deep Learning is the inferred idea of AI with an algorithm inspired from the human neuron structure, called the artificial neural network. A hierarchical learning (deep structured learning) algorithm is valuable for classification and prediction issues and fit for content interpretation with understanding and reasoning and, furthermore, ready to transfer and apply the result derived from the learning model to the latest situation. Deep Learning uses algorithms, such as feed forward and back propagation, to describe the best results for applications using AI and developing a Deep Learning based neural network intrusion detection system [4].

13.2   Literature Review

Anand T, Rahul M, S, Vasantha Kumari, M, Narendran G (2020), Intrusion Detection Using Deep Learning Algorithms, Journal of Engineering, Science In this paper, neural system models anticipate the assault data and indicate the type of attack based on real time network intrusion datasets. The model ought to be more productive regarding time to predict the assault and, furthermore, be successful over the prediction of thee type of attack. In the future, the system can be additionally extended by refining the network model to moderate the systematic assault and upgrade network security.

Farah Barika Ktata (2020), https://doi.org/10.1007/s42452-020-2414-z, A Deep Learning-Based Multi-agent System for Intrusion Detection In this paper, IDS is developing various operators that embedded three algorithms: the explicit autoencoder (AE), multilayer perceptron (MLP), and k-nearest neighbor (K-NN). The autoencoder helps to reduce the feature of the task and with MLP and K-NN, the classification process can be done. The KDD 99 benchmark dataset is used to survey the proposed model. In future work, we can apply this arrangement in real network traffic (i.e., online IDS) to Benefit from other AI by including more specialist classifiers to grow the accuracy of interruption identification and extend our IDS so it will, in general, be used with Cloud Computing, fog computing, and the Internet of Things for security reasons.

Prasanna Kottapalle (2020), A CNN-LSTM Model for Intrusion Detection System from High Dimensional Data, *Journal of Information and Computational Science, ISSN: 1548-7741* In this paper, the Convolution Neural Network algorithm and LSTM algorithm are merged on the system. The experiment shows that this model will expand the accuracy of identification of human interference and boost the productivity of the detection strategy for human intrusion. It is likewise logical that the anticipated portrayal beat SVM, DBN, and CNN with the CNN-LSTM model [5].

Khraisat *et al.* Cybersecurity (2019), https://doi.org/10.1186/s42400-019-0038-7, Survey of Intrusion Detection Systems: Techniques, Datasets, and Challenges In this paper, an overview of IDS techniques, categories, and innovations are introduced. A few AI strategies that have been proposed to recognize zero-day network intrusion are looked into. Regardless, such approaches may have the issue of creating and invigorating the information about new threats and output frequency of wrong alarms or poor precision. Furthermore, a major challenge in this area of research is developing IDSs equipped for defeating avoidance methods.

Yang Jia, Meng Wang, Yagang Wang (2018), Network Intrusion Detection Algorithm Based on Deep Neural Network, ISSN 1751-8709, doi: 10.1049/iet-ifs. 2018. 5258, Journals the Institute of Engineering and Technology In this paper, a DNN-based IDS is built and the analysis results show that capacity and execution of the NDNN-based IDS are better than strategies dependent on conventional AI strategy. Utilizing the NDNN organize model to do intrusion detection is feasible, especially in this multi-feature dataset. In the future, high location rate on datasets does not mean comparative execution in actual detection. In this way, more network simulation experiments need to be done.

Santhosh Parampottupadam (2018), Cloud-based Real-time Network Intrusion Detection Using Deep Learning In this paper, an evaluation study was finished using the benchmark NSL-KDD dataset which includes typical traffic records similarly as interferences gathered into four classes: U2R, Probe, R2L, and U2R. The evaluation study analyzed Deep Learning models built using H2O and Deep Learning4J libraries, with other typically used AI models, for instance, SVM, Random Forest, Naïve Bayes, and Logistic Regression. The future work is to improve the estimation of detection rates of multinomial classifiers and headings using datasets with other real time network intrusion traffic [6].

Kaushik Roy and Albert Esterline (2020), Using a Long Short-Term Memory Recurrent Neural Network (LSTM-RNN) to Classify Network Attacks, doi:10.3390/info11050243 In this paper, the NSL-KDD dataset for both binary and multi-class classifications have different classifiers. They considered SVM, Random Forest, and the LSTM-RNN model. In this paper, the proposed model delivered a higher accuracy rate for binary classification. The LSTM-RNN acquired higher precision than the SVM in binary classification. Future works will extend the experiment for contrasting the performance of LSTM_RNN with GA to those of other Deep Learning approaches, such as CNN-LSTM, STL, and a deep belief network (DBN) on the latest datasets.

Leila Mohammadpour, Teck Chaw Ling (2018), A Convolutional Neural Network for Network Intrusion Detection System, ISBN 978-4-9905448-8-1, Proceedings of the APAN In this paper, the feasibility of CNN in the field of NIDS to detect the network anomaly is discussed. The NSL KDD approach was used to detect the network anomaly. The future work would be to evaluate a real-time dataset to dynamically detect various forms of potential attacks.

Jin Kim, Nara Shin, Seung Yeon Jo and Sang Hyun Kim (2017), Method of Intrusion Detection using Deep Neural Network, Published in IEEE In this paper, an AI Intrusion Detection System utilizing a deep neural network (DNN) was researched and tried with the KDD Cup 99 dataset because of ever-developing system assaults [7].

13.3   Incursion In Cloud

An endeavor to bargain the confidentiality, integrity, or accessibility of a network or system is known as an incursion (intrusion). In this segment, significant classes of intrusion that regularly influence the cloud are depicted. This is trailed by an introduction of different attacks on the cloud characterized for the cloud's model.

13.3.1   Denial of Service (DoS) Attack

The attacker utilizes bots (zombies) for flooding a framework with an enormous number of bundles to deliver the accessible assets inaccessible. In this way, the services for the time being are not accessible on the Internet. As per some vulnerability specialists, an intruder can influence more clients by launching a DoS attack on the cloud [8].

13.3.2   Insider Attack

An insider attack is characterized as a current employee or associate of the cloud specialist co-op which has advantaged access and the position to perform changes in the cloud environment. Insider attacks are sorted out as they have information about the client and providers. This is lethal because the greatest number of attacks can be executed from inside and an intruder can easily evade recognition without appropriate controllers.

Figure 13.2 Intrusion in cloud.

A DoS attack by an insider was propelled on the Amazon Elastic Compute Cloud (EC2) and cloud clients' confidentiality was penetrated in this attack shown in Figure 13.2.

13.3.3   User To Root (U2R) Attack

In this attack, the attacker gets access to the credential information of the legitimate end user and, afterward, exploits system accountability to get the root benefits. In the cloud, the intruder first gets to an occurrence and exploits its weaknesses to accomplish root benefits of a virtual machine or host. By this attack, the integrity of the cloud is being abused [9].

13.3.4   Port Scanning

Port scanning is utilized by the intruder to get data about open, shut, filtered, and unfiltered ports. The intruder at that point utilizes this data to dispatch attacks on open ports. Various strategies are utilized to perform port scanning. This attack focuses on the confidentiality and integrity of the cloud.

13.4   Intrusion Detection System

An interruption can be characterized as an unapproved activity that can cause pulverization to a data framework. This suggests any assault that could represent an expected threat to data CIA will be seen as an interference or interruption. For example, an action that would make the system organizations lethargic to the authentic client is seen as an interruption. An IDS is a product or equipment apparatus that perceives malicious activities on PC framework in order to keep up security [10]. The objective of an intrusion system is to distinguish various types of malevolent network traffic and PC use that cannot be perceived by a conventional firewall. It is essential to create better security from threats that deal with the CIA of system. The intrusion system framework is comprehensively divided in three types: a Signature Intrusion Detection System (SIDS), an Anomaly Intrusion Detection System (AIDS), and a Deep Learning Intrusion Detection System.

13.4.1   Signature-Based Intrusion Detection System (SIDS)

The techniques of SIDS depend on similar patterns matching strategies to detect a familiar assault. In SIDS, pattern matching strategies are utilized to locate a past interruption. As such, an interruption signature matches with the characteristic of a past interruption starting at the present in the mark database and an alert sign is activated. For SIDS, host's logs are evaluated to find progressions of requests or activities recently perceived as malware. SIDS have similarly been named in the composition as Knowledge-Based Detection. The traditional way to deal with this was by managing SIDS review network packets and attempting to coordinate against a database of signature. However, procedures cannot recognize attacks that length of network packets. As current malware is more prevalent in the present day, it may be critical to remove signature data over different packets. This requires the IDS to audit the content of prior packages. Concerning making a signature for SIDS, for the most part, there have been various techniques where a signature is made as a state machine, formal language string pattern, or semantic conditions. The expanding pace of zero-day attacks has delivered SIDS strategies continuously that are less powerful due to the fact that no earlier signature is present on such assaults [11]. Polymorphic varieties of malware and the increasing measure of focused assaults can additionally underline this conventional paradigm. A possible answer for this issue is the use of AIDS procedures, which work by profiling what is satisfactory conduct, as opposed to what is anomalous, as mentioned in the following segment.

13.4.2   Anomaly-Based Intrusion Detection System (AIDS)

An anomaly-based computer system utilizes the concepts of AI, Deep Learning, and statistical based strategies and improves the overall performance of IDS. Any variance between the analyzed conduct and the model is viewed as an anomaly, which can be deciphered as an interruption. The assumption for this gathering of strategies is that malevolent strategies contrast in distinction to normal client conduct. Activities of strange clients which are not similar to standard practices are named intrusions. Improvement of the anomaly system involves two stages: a preparation stage and a test stage. In the preparation stage, a heavy transit profile is used to gain capability illustrative of typical conduct and, a short time later, in the test stage, another informational collection is used to build up the framework's ability to summarize effectively unnoticeable interferences or intrusion. Helps can be ordered into various classifications dependent on the strategy utilized for preparation (training), for example, statistical based, knowledge based, or AI (machine learning) based. The primary favorable position of AIDS is the capacity to distinguish zero-day attacks because of the way perceiving irregular client movement does not depend on a

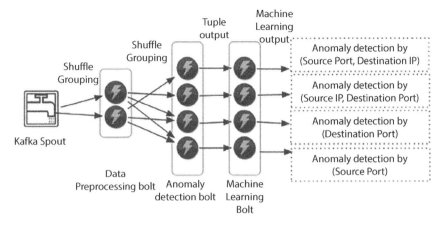

Figure 13.3 Real time network anomaly detection system.

signature database. Helps triggers a peril signal when the analyzed conduct varies from standard conduct. Moreover, AIDS has different advantages. To begin with, they have the ability to find interior malicious activities [12]. On the off chance that an interloper begins making exchanges in a taken record anonymously in the ordinary client movement, it makes an alert. Also, it is incredibly hard for a cyber criminal to see what a normal customer conducts without being prepared, as the framework is worked from altered profiles shown in Figure 13.3.

13.4.3   Intrusion Detection System Using Deep Learning

Various analysts have explored Deep Learning for intrusion detection, utilized deep neural networks (DNN) and basic data cleaning, and duplicated elimination features to preprocess the data along with training data conversion. Since the KDDcup99 dataset contains entire numbers and coasting values, they changed over all the data into string types to restrict data misfortune. The developers utilized 0.10% of the altered information for the preparing model and achieved 0.99% precision with an 8% negative (false) alert rate [13].

Saxe and Berlin used more than four hundred thousand programming pairs for building a Deep Learning model including four layers. The chief layer gets the preparing layer, the second is arranging the information dependent on labelled data and unlabeled data, and the last layer is for classification. Analysts accomplished 0.95% precision. Data cleaning was not referenced and this is significant given that the KDDcup99 dataset contains over 0.70% replicated records.

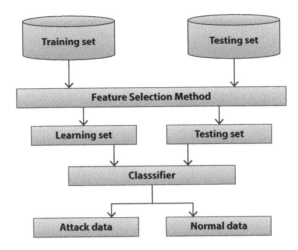

Figure 13.4 Deep learning based intrusion detection system.

The KDDcup99 dataset contains real-time network attack values and the information of five diverse network attacks were added to it. In the primary module in this chapter, the real-time KDDcup99 dataset is preprocessed and changed over from a raw set to a clean dataset by utilizing normalisation functions and numerializing the dataset values. The clean dataset is taken care of to a feature selection and extraction modules which divide the dataset into preparation/training model data and testing model data. The prepared dataset is taken care to the Deep Learning model at the second stage. In the second stage, the Deep Learning models are characterized to process the testing model information shown in Figure 13.4.

Deep Learning modules contain CNN, RNN, and MLP network models that proceduralize the information and each model gives distinctive accuracy over the input model data. A Neural Network Model with high accuracy is chosen to predict the type of threat over the flask server. The flask server model acquires the best system model to indicate the type of attack over the given log records from the client input log files [14].

13.5   Types of IDS in Cloud

13.5.1   Host Intrusion Detection System

The Host Interruption Detection System (HIDS) acts as a specific server (host) and monitors the activity of just that system. HIDS methods

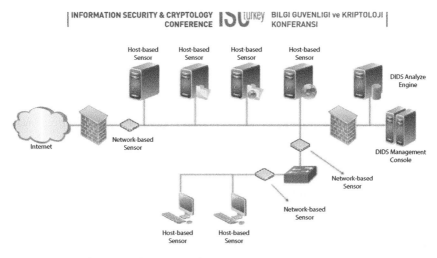

Figure 13.5 Architecture of IDS in cloud.

concerning cloud, for the most part, can be partitioned into three fundamental organization based groupings. HIDS, inside the VM for monitoring, can be deployed in the host OS or in a different, visitor OS shown in Figure 13.5.

HIDS gathers data from explicit hosts and examines for distinguishing meddling occasions. Here, the study depends on host-bound data, for example, the operating system, clients, and applications.

13.5.2 Network Based Intrusion Detection System

It is considered as network traffic and observes numerous nodes to recognize interlopers. When an assault is identified, unexpected action is detected and an executive will receive an alarm. NIDS holds traffic of a whole network which investigates potential interruptions like DOS attacks, port scanning, and so forth by using IP and transport layer headers of caught network packets. This system utilizes anomaly based and signature based detection methods for recognizing interruptions. Numerous hosts are sent in networks that can be secured from interlopers dependent on using legitimate NIDS.

13.5.3 Distributed Based Intrusion Detection System

Distributed Based Intrusion Detection Systems comprise of a large network having numerous IDS and each and every one can communicate with another or centralized server. To counter DoS and DDoS attacks in the cloud, we have proposed and simulated an interruption detection system. The IDS has four segments each assuming a particular job. This protects the system from a solitary purpose of loss. Be that as it may, it utilizes a signature based discovery method because unseen (unknown) attacks are not identified.

DIDS is a blend of numerous IDS, for example, HIDS and NIDS for observing traffic from bigger networks and, furthermore, keeping away interlopers. DIDS has, mostly, two components, for example, (a) a detection segment is utilized to monitor the system or subnet or (b) a correlation manager assembles data from different a IDS and is utilized to produce more level alarms for interloper awareness [15].

13.6   Model of Deep Learning

The Deep Learning models of CNN, RNN, and MLP are the system models chosen to prepare the preprocessed information and the obtained model is contrasted among them to get the best expectation model. CNN, RNN, and MLP neural network models have a prevalent structure of 3-layers which incorporates input, hidden, and output layers. These system models give better execution over the tabular information. Keras and Tensor Flow packages are utilized as frontend and backend systems separately for Deep Learning models. CNN, RNN, and MLP models give the preparation/training models output information which can be utilized to predict attack information and would have the option to indicate that kind of attack.

13.6.1   ConvNet Model

The Convolutional Neural Network Model, CNN Model, or ConvNet Model, involves a 3-layer structure that incorporates an input data layer, hidden layer as a convolutional pooling layer with an implement (activation) layer, and an output layer.

The CNN is a variety of the neural network, where it will likely learn appropriate component representation of the input data. A CNN has two primary contrasts with MLPs, including weight sharing and pooling. Each layer of CNN can be made out of numerous convolution pieces which are utilized to produce a distinctive component map. Every region of neighboring neurons is associated with a neuron feature map of the next layer. Moreover, to create the feature map, all the spatial areas of the input share the part. After some convolution and pooling layers, one or various full associated layers are utilized for classification.

CNN Neural structure works on CSV information and can give better execution over classification and prediction issues. The CNN Model is implemented in a consecutive manner which incorporates the way toward the input layer with a one-dimensional convolutional layer, followed by installing the thick layer with the actuation function as softmax. The CNN model is incorporated and compiled by the enhanced agent adam class with a loss count of categorical cross entropy.

13.6.2   Recurrent Neural Network

A Recurrent Neural Network, or RNN Model, has a successive model structure like the CNN Model. A Recurrent Neural Network, or RNN, seems as though a classical neural system is an artificial neural network. In a conventional neural system, the model delivers the output by increasing the contribution of the weight and activation function. In the RNN, data can spread in the two bearings, including from deep layers to the main layers. In this, they are nearer to the true functioning of the sensory system which is not one way. These networks have recurring connectivity, as in they keep data in memory. In principle, RNN should transmit data on schedule.

The RNN Model involves a 3-layer structure that incorporates an input layer, moderate hidden layer with forward and backward propagation functionalities, and an output layer. The RNN Model is more compatible with tabular data and gives better outcomes over classification and prediction problems and regression prediction issues. RNN Models have an inward structure of Long Short-Term Memory (LSTM) usefulness which is proficient to deal with sequence prediction issues.

The RNN Model has been actualized in a successive manner which incorporates the way towards embedding the input layer with and LSTM layer, followed by coordinating the hidden layer with the activation function as a sigmoid. On continuation with the installing procedure, a thick layer is included with the hidden layer. The RNN Model is aggregated by utilizing the optimizing agent adam class with the loss computation of cross entropy.

13.6.3   Multi-Layer Perception Model

The Multi-Layer Perception Model or MLP Model, also called a regression classifier, empowers changes in the input data in a non-linear structure. This non-linear structure information is taken care of to the immediate hidden layer of perception and from it they prepare the output layer. The immediate hidden layer utilizes non-linear actuation functions to perform regression predictions and classification issues [16]. The MLP model is viewed as the widespread approximator since its interior structure depends on XOR functions. MLP, also called a Feed-Forward Neural Network, utilizes a regulated (supervised) learning method called back-propagation for consistent functions and preparing the model shown in Figure 13.6.

The deep neural network is a feed-forward neural network and is a super set of multilayer perception. This perceptron methodology relies

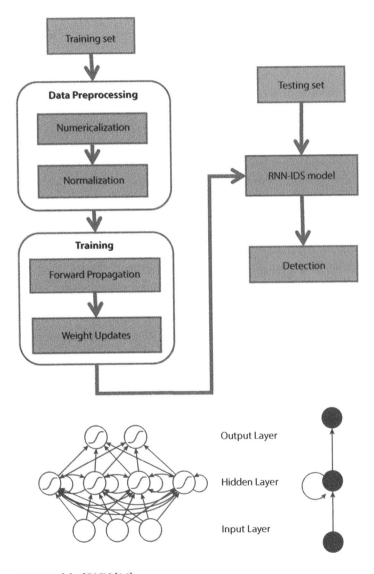

Figure 13.6 Model of RNN [16].

upon a backpropagation algorithm which works with the predicated error propagation technique. It involves couple phases: a Feed Forward phase and a Back Forward phase.

In the first step, data propagates over the network to finally get the output and contrasts it with the real values to get the error. Afterward, to be minimized, the error is backpropagated to the previous layer, and then the loads are balanced in a like manner. This procedure is repeated until the error is under a predetermined threshold.

13.7   KDD Dataset

To analyze the performance of an anomaly IDS, we use the KDDcup99 Dataset. It was first introduced in 1999 in academic research, specially used for a concept in machine learning, IDS. The KDD cup99 dataset is generally used to analyze the IDS algorithm. It has named assault tests that are learned by impartial observing. The assault can be of any one of the types discussed in the above section 4. The names of attacks are DoS, U2R, R2L, and probe. These datasets are viewed as appropriate to predict the assault type and its data values. Generally, the analysis of the algorithm depends on their properties. The properties are ordered under 5 variables:

7.1 Essential Highlights: In this, we considered the type of connection established was TCP/IP or UDP.

7.2 Repetitive Highlights: In this, we considered the duration, type of protocol, flag, land, source byte, destination byte, etc.

7.3 Content Highlights: In this, we considered the number of logins, number of failed logins, number of files created, number of access files, whether login was as a host or guest, etc.

7.4 Traffic Hightlights: In this, we considered the features of transmission between source to destination like sender error rate, receiver error rate, same server rate, different server rate, server count, etc.

7.5 Host Highlights: In this, we considered the features related to host like destination host count, destination host server count, destination host same server rate, etc.

13.8   Evaluation

We take an experiment for estimating the presentation of our model. To assess the presentation of a model, there are a number of ways to calculate the performance metrics, for example, precision-rate, efficiency-rate, false-positive rate (FPR), and true-positive rate (TPR), True Positive (TP), False Negative (F N), False Positive (FP), and True Negative (TN).

- True-positive Rate is the ratio when an actual attack happens and intrusion detection gives an alarm.

$$TPR = \frac{TP}{(TP + FN)}$$

- True-negative Rate is the ratio when no attack happens and intrusion detection gives no alarm.

$$TNR = \frac{TN}{(TN + FP)}$$

- False-positive Rate is the ratio when no attack happens and intrusion detection gives an alarm.

$$FPR = \frac{FP}{(FP + TN)}$$

FP : False Positive
TN : True Negative
- False-negative rate is the ratio when an actual attack happens, but intrusion detection gives no alarm.

$$FNR = \frac{FN}{(FN + TP)}$$

- Efficiency rate is the ratio between the actual intrusion detection versus overall detection.

$$Effieciency - rate = \frac{(TP + TN)}{(TP + TN + FP + FN)}$$

- Precision rate is the ratio between the truly strange packets versus absolute packets that were set apart as intrusions.

$$Precision - rate = \frac{TP}{TP + FP}$$

- F-Score is the uniform mean of the true-positive rate and precision.

$$F - Score = \frac{2 * Precision - Rate * TPR}{TPR + Precision - Rate}$$

13.9 Conclusion

The principle target of our task work is to utilize the Deep Learning model to predict the sort of assaults dependent on the real time network assault dataset values and, furthermore, indicate the assault data which helps in rapid avoidance of the classified network assault type. The algorithm was based on 3 models: CNN, RNN, MLP. With the help of these models, inputs can calculate the true positive rate, false positive rate, true negative rate, false negative rate, precision rate, efficiency rate, and F-score. On the basis of their inputs, in these terms, we get that which model is more reliable. In this chapter, we are reviewing the algorithm, but in future we will practically explain the algorithm with the help of programming language.

References

1. Abhishek Kumar & Jyotir Moy Chatterjee & Pramod Singh Rathore, 2020. "Smartphone Confrontational Applications and Security Issues," International Journal of Risk and Contingency Management (IJRCM), IGI Global, vol. 9(2), pages 1-18, April.
2. Anand T, Rahul M, S, Vasantha Kumari, M, Narendran G (2020), Intrusion Detection Using Deep Learning Algorithms, journal of engineering, science.
3. Bhargava, N., Bhargava, R., Rathore, P. S., & Kumar, A. (2020). Texture Recognition Using Gabor Filter for Extracting Feature Vectors With the Regression Mining Algorithm. International Journal of Risk and Contingency Management (IJRCM), 9(3), 31-44. doi:10.4018/IJRCM.2020070103.
4. Farah Barika Ktata (2020), A deep learning-based multi-agent system for intrusion detection, https://doi.org/10.1007/s42452-020-2414-z.
5. Jin Kim, Nara Shin, Seung Yeon Jo and Sang Hyun Kim (2017), Method of Intrusion Detection using Deep Neural Network, Published in IEEE.
6. Kaushik Roy and Albert Esterline (2020), Using a Long Short-Term Memory Recurrent Neural Network (LSTM-RNN) to Classify Network Attacks, doi:10.3390/info11050243.

7. Khraisat *et al.* Cybersecurity (2019), Survey of intrusion detection systems: techniques, datasets and challenges, https://doi.org/10.1186/s42400-0 19-0038-7.

8. Kumar, A., Chatterjee, J. M., & Díaz, V. G. (2020). A novel hybrid approach of svm combined with nlp and probabilistic neural network for email phishing. International Journal of Electrical and Computer Engineering, 10(1), 486.

9. Leila Mohammadpour, Teck Chaw Ling (2018), A Convolutional Neural Network for Network Intrusion Detection System, ISBN 978-4-9905448-8-1, Proceedings of the APAN.

10. N. Bhargava, S. Dayma, A. Kumar and P. Singh, "An approach for classification using simple CART algorithm in WEKA," 2017 11th International Conference on Intelligent Systems and Control (ISCO), Coimbatore, 2017, pp. 212-216, doi: 10.1109/ISCO.2017.7855983.

11. Naveen Kumar, Prakarti Triwedi, Pramod Singh Rathore, "An Adaptive Approach for image adaptive watermarking using Elliptical curve cryptography (ECC)", First International Conference on Information Technology and Knowledge Management pp. 89–92, ISSN 2300-5963 ACSIS, Vol. 14 DOI: 10.15439/2018KM19.

12. Prasanna Kottapalle (2020), A CNN-LSTM Model for Intrusion Detection system from High Dimensional Data, Journal of Information and Computational Science, ISSN: 1548-7741.

13. Rathore, P.S., Chatterjee, J.M., Kumar, A. *et al.* Energy-efficient cluster head selection through relay approach for WSN. J Supercomput (2021). https://doi.org/10.1007/s11227- 020-03593-4.

14. Santhosh Parampottupadam (2018), Cloud-based Real-time Network Intrusion Detection Using Deep Learning.

15. Singh Rathore, P., Kumar, A., & Gracia-Diaz, V. (2020). A Holistic Methodology for Improved RFID Network Lifetime by Advanced Cluster Head Selection using Dragonfly Algorithm. International Journal Of Interactive Multimedia And Artificial Intelligence, 6 (Regular Issue), 8. http://doi.org/10.9781/ijimai.2020.05.003.

16. Yang Jia, Meng Wang, Yagang Wang (2018), Network intrusion detection algorithm based on deep neural network, ISSN 1751-8709, doi: 10.1049/iet-ifs. 2018. 5258, journals the institute of Engineering and Technology.

14

Applications of Deep Learning in Cloud Security

Disha Shrmali* and Shweta Sharma

MDS University, Ajmer, India

Abstract

Image processing, an enormous amount of file sharing and storage, data backup, testing and development of large amounts of data, etc. requires a well rooted framework to enjoy the benefits of memory, space, and computational time. Parallelly, it also requires us to embrace affordable answers to supplant conventional systems. These necessities lead us to utilize cloud computing to meet broad information prerequisites. In the meantime, this standpoint gives quick access to on-request help with high accessibility and adaptability. Consequently, working with cloud benefits rather than in-house applications would, without a doubt, pitch into numerous business, educational, and medicinal services, just as security based authoritative redistribution calculations to any outer gathering, in this way reducing undesirable activity-based costs. All things considered, solid information protecting against both untrusted mists and vindictive clients is requiring deflecting unapproved information revelations. In today's situation, mysterious structures are initiated to empower clients to store and proceduralize their information utilizing distributed computing. These structures are commonly developed utilizing cryptosystems, appropriated frameworks, and some of the time, a mix of both. To be explicit, homomorphic cryptosystems, attribute-based encryption (ABE), Service-Oriented Architecture (SOA), Secure Multi-party Computation (SMC), and Secret Share Schemes (SSS) are the significant security systems being the most widely used. The principal issues being looked at during the time spent on a gigantic information investigation over the cloud utilizing these methods are the computational expenses related with all handling errands, violations from other users of the cloud, insufficient security of Internet channels, and absence of accessibility to resources. In this chapter, we will learn about all the security issues being faced by users in the Cloud Computing environment and about a new outlook to resolve these issues with the help of deep

---

Corresponding author: dishashrimali799@gmail.com

Pramod Singh Rathore, Vishal Dutt, Rashmi Agrawal, Satya Murthy Sasubilli, and Srinivasa Rao Swarna (eds.) Deep Learning Approaches to Cloud Security, (225–256) © 2022 Scrivener Publishing LLC

methods to protect data dispensation in the cloud atmosphere. We will also look towards the live projects and applications that have been developed to provide validation towards our outlook, which includes their working, the technologies are used to penetrate better performance and cost efficiency for users, the advantages they are providing to enhance efficiency in data processing, and the further disadvantages users might be facing while working with it.

Keywords: Cloud computing, deep learning, firewall, web application firewall (WAF)

14.1   Introduction

Cloud Computing Security and Issues

Honouring the observations of the Cloud Security Alliance (CSA), it is not a big surprise to say that above 70% of world's organisations use cloud services for their institutions to work. This is because they have knowledge of its major attributes like lower fixed costs, higher flexibility, automatic software updates, increased collaboration, and the freedom to work from anywhere. But, as history shows, nothing is perfect and there is a list of issues being faced by organisations working under the shadow of a cloud. In recent events, a report published over cloud security suggested that about 60-70 percent of companies are pressured to improve public cloud security. This pressure is because of the existing fact that there are many attackers around the world who have an enormous number of tactics to compromise privacy. Cloud services might have created many new methods to detect malware, yet there are many companies who still don't want to take steps towards the cloud without a proper roadmap. Let's have a look over some of the security issues faced in the cloud environment [1].

14.1.1   Data Breaches

Cloud Computing in the history of computer science is a very recent development, but when we talk about cyber-attacks, we know that people have been doing these practices for years. The main question creating glitch in the matrix is whether the cloud is the only premise that is fundamentally less safe if we choose to save our data on it rather than on our own premises. The Ponemon Institute presented a report concluding that over 50% of professional security observers doubt the efficiency of security applications or hardware being used by organisations at present. To provide validation to the study, they created nine frameworks and each framework was evaluated by reviewing its detailed reports and the con-clusion stated that those businesses that acquire cloud assistance are three times more likely to be affected by malicious activities then those who do not. It was transparent to state that Cloud Computing follows certain types of ambiguities that make its environment prone to vulnerabilities [2].

14.1.2   Accounts Hijacking

The existence of the cloud in many organisations has activated an enormous number of issues, especially in terms of account hijacking. From all around the world, attackers can easily access yours and your employee's information to access your accounts so that they could easily and discreetly access the essential credentials of the company and misuse them for faulty purposes. They can do that by using bug scripting or randomly reused passwords [3].

Amazon faced a cross-site scripting bug in April 2010 targeting fabrica-tion of customer credentials. Phishing, key logging, and buffer overflow are some other account hijacking thefts. In the middle of the attack, attackers can access the employees or user key. Because of this act, cloud-oriented systems verify individual devices without letting the user log in to the device each time during update and sync [4].

14.1.3   Insider Threat

When we own a company, we consider it our family. In such cases, we do not see the attacker living inside the premise, but that does not mean they do not exist. Your employee can easily misuse their privileges and access to the company's files and services and can misuse them to create a threat to the company. Sometimes even the employees do not have intentions to threaten the organisation, they just happen to do certain things that create a threat. A survey conducted by Imperva suggests that an insider threat is nothing, but the harm caused by an insider intentionally or unintentionally by getting access over cloud of organisation and accessing the data might create chaos for the organisation. It suggests that some steps can be taken by the organisation to prevent the situation like insider threats, verified partnerships, access controls, preauthorisation of initiatives taken to safe-guard the cloud, etc. [5].

14.1.3.1   Malware Injection

When a hacker or attacker tries to eaves drop in cloud, he injects a script of codes in the system. These codes, once validated, start acting like software as a service (SAAS) for the system and seem to be just as the part of the system. Once this code is executed and the cloud becomes dependent on it, the attacker can easily get access to the data and cause any harm to the company or system [6]. This process in known as a malware injection. According to a report from East Carolina University, a malware injection attack is the most dangerous attack for Cloud Computing systems.

14.1.3.2   Abuse of Cloud Services

Every day, with each passing minute, the cloud industry is expanding its branches in every sector of life. It has made it possible for many organisations to store or transfer large amounts of data in much less time and in a very cost friendly manner. But, due to the same reason, many attackers and data breaches also find a way to host unworthy malware, illegal software, etc. This affects both the service provider the clients of the service provider.

Risks such as channelling of private videos, pictures, data, etc. can be considered as some factors whose occurrence can cause some legal issues against copyright law. According to the damage, the fine amount can reach up to $250,000 as well. To minimise your risk factor, you must keep a list of usage and the types of data your employees are storing.

14.1.3.3   Insecure APIs

For those users who like to customise their cloud according to their needs, the best application they can get is API. An application programming interface is an application that provides a platform for the user to create a customised cloud according to their needs. Having an API has many advantages of its own, yet it has some disadvantages as well. The increase in the technological sector of API is directly proportional to no increase in risk factors. API provides a number of tools to the user to amalgamate their application according to other software's. The best example is You-Tube. The security risks in such applications increase when transition of data occurs within different applications [7].

14.1.3.4   Denial of Service Attacks

There are many types of attack where an intruder in any way can harm your data. A denial of service attack is a bit different from normal attacks where the attacker tries to barge into the system and extract information or corrupt the system. They disable the verified user to get direct access to the Website or server. Due to this, the user thinks that the website is not giving a follow up to his request, which results in client loss for the business owner. If the attacker has good knowledge, he can also use DOS as an imitation image to breach security applications, such as a web application firewall.

14.1.3.5   Insufficient Due Diligence

When an organisation is not 100% considerate and goal oriented, it might make certain decisions without proper planning and research and without verifying the authentication of the provider. In such cases, a great amount of loss in the company might occur. Due diligence becomes a great risk for a company when it suddenly makes the decision to switch to the cloud without doing any presurveying. This is not a technical error, but just a glitch in proper decision making. Companies like PCI, FERPA, etc., whose data follow a set of regulatory laws, need to pay major attention to it.

14.1.3.6   Shared Vulnerabilities

When we talk about cloud security, we are narrating an open gateway being shared in between the service provider and our client. In such a situation, it is obvious from the client's side to take some very strict actions to save their data from any kind of malicious activity. Though many major cloud benefactors like Google, Microsoft, Amazon, etc. have their own security systems to secure their data from breach, this totally depends on an individual client to secure its data. As it has been suggested in the article, one must follow some security procedures, like there should be a limitation of access to data, proper authentication and a validation procedure followed while storing data, one must encrypt their password with a proper ke , one must imply limitations to both his files and devices, etc. [8].

14.1.3.7   Data Loss

There is usually a large amount of data present in a cloud environment. As we know that larger amount of data increases the probability of malicious attack, data breach, or loss of data. Your business will be dragged towards the route of destruction once its essential information is leaked. The best example in the history of data loss happened with Amazon. The organisation faced a permanent loss of data which was never recovered. The next one on the list is Google. Lightning struck Google's power grid four times. It is a very important task to secure your data. Before accepting any data request, you must revaluate the backup procedure of your providers.

14.2   Deep Learning Methods for Cloud Cyber Security

In this chapter, we will learn about diverse Deep Learning approaches utilized in cloud cyber security with references to significant strategical examples accommodated in every procedure.

14.2.1   Deep Belief Networks

DBNs work as a class of DNNs, which are made out of various nodes of shrouded sets with interrelations within the nodes, although in-between the units inside layers. DBNs are prepared in an unverified environment. Usually, they are made by altering load concealed layers separately to, again, make the inputs.

14.2.1.1   *Deep Autoencoders*

Autoencoders are basically an unadministrated class of neural networks where inputs are provided to the network in the form of a vector and attempt coordinating received outputs equivalent to that vector. To make the information portrayed sequentially dimensional, one must take the input, change dimensionality, and reproduce the information. These kinds of neural systems comprise of unimaginably adaptable features because of the fact that they acquire the knowledge of compressed data indoctrination in an unverified form [9]. Furthermore, they can train layers one by one, due to which a powerful model is manufactured with less energy consumed. In this process, when the moment arises

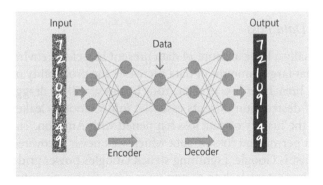

Figure 14.1 Deep autoencoder.

when input and output layers have higher dimensionality which the system utilised for data encoding, the autoencoder intends to evacuate commotion and strengthen itself by training the autoencoder to recreate contribution with an uproarious form of input, known as a denoising autoencoder.

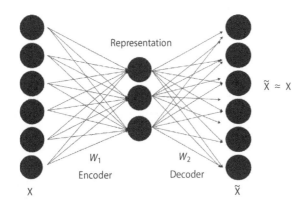

Figure 14.2 Denoising autoencoder.

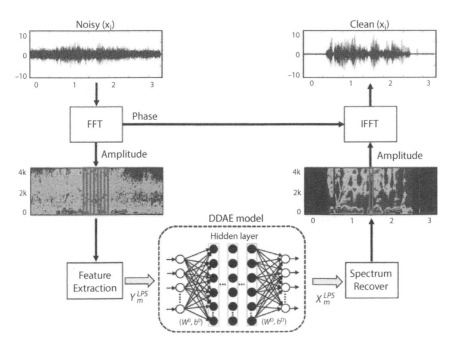

Figure 14.3 Deep autoencoder and denoising autoencoder.

Utilizing various blocks of autoencoders prepared to arrange and step by step pack the information in, increasing the format, is known as a stacked autoencoder. The figures given below are basically finished stack autoencoders. They are framed by making an autoencoder, as shown in Figure 14.2. At that point, the autoencoder in Figure 14.3 is constructed utilizing the yields of Figure 14.2; these are consolidated and a classified layer is included. Like ordinary autoencoders, we can also stack denoising autoencoders [10] shown in Figures 14.1, 14.2 and 14.3.

14.2.1.2 Restricted Boltzmann Machines

Restricted Boltzmann Machines (RBMs) are dual-layer, bi-parted, directionless graphical models where Data can brook in between the two bearers so it can provide an architectural form to DBNs. RBMs are single-handedly just like autoencoders and they can be prepared one layer after another. The 1st layer is the input layer, followed by the 2nd layer that is the concealed layer. Interconnections are formed in between the layers having similar nodes, but each hub in the 1st layer is associated with each hub in the shrouded layer [11].

Commonly, these layers are confined to carrying binary units. The network is restricted to limit the vitality-rate of the functions that enumerate the similarities in the model, obtaining a great part of arithmetic from a numerical mechanism. The major point in the training the model is to determine function and the concealed position that binds the vitality of the structure. Also, RBMs allocate probabilities, not rock-hard values. In any case, for other models, the output can be utilized as a highlighting feature. The structure of the model is equipped by accepting binary input data and pushing it forward through different layers. At the same time, it pushes the information

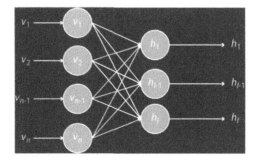

Figure 14.4 Restricted boltzmann machine.

in reverse through the layers of model to reproduce the information. At this point, the vitality of the framework is determined and is utilized to refresh the pages [12]. This this must be done continually until the model converges shown in Figure 14.4.

14.2.1.3   DBNs, RBMs, or Deep Autoencoders Coupled with Classification Layers

Grouping layers, when combined with RBMs and autoencoders, can be linked to produce out an arrangement utilizing completely associated branches. The branches prepared by inducing Deep Learning are used as feature extractors and they create contributions to associated branches that are prepared with the help of back propagation. In contradiction to RBM or autoencoder layers, these branches need labels to fix. These types of schemes have designated achievements in numerous apps, considering auditory demonstration and modelling and speech and image recognition.

14.2.1.4   Recurrent Neural Networks

A recurrent neural network (RNN) is an upgrade to the traditional neural network, where data inputs can be accepted to a fixed size so that it can handle mutable spans of input sequences. RNN accepts and processes inputs for a single component by using the output of enclosed units as an additive input to the next component. Hence, RNNs can discourse language, time series, and speech problems.

Commonly, RNNs are harder to prepare because the gradients can undoubtedly vanish or explore. However, progressions in preparing and engineering have created numerous RNNs that are simpler to prepare. As a result, RNNs have indicated achievement word analysis in sentence forecast, talking acknowledgment, language paraphrasing, picture captioning, and other period relevant forecast tasks. The concealed units of an RNN are fit for keeping up a state vector that stores past memories in a queue form. The type of RNN node determines the length of memory. The RNN is fit for learning at more extended terms of the conditions if the memory is high. RNNs ignore issues that need longer life memories with the help of LSTM units. LSTM units consist of an architecture considered as memory cells to analyse data and interfaces oneself with that data in the next step. The memory cell expands its estimation cells with new info and a forgetting gate that stores more data reliant upon what is required shown in Figure 14.5 [13].

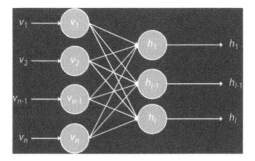

Figure 14.5 Recurrent neural networks.

14.2.1.5   Convolutional Neural Networks

A CNN is a neural system whose only interest is to evaluate input stored in an array. Information which a model provides is generally in the form of shading or a grayscale picture, or we can say a two-D matrix of pixels. For handling 2D varieties of pictures or spectrograms of sound, CNNs are regularly utilized. Similarly, they use three-dimensional displays. Use of a 1D array is less regular yet very important. Where there is a longitudinal or temporary request, CNNs are demanded [14].

The engineering of Convolutional Neural Networks occupies three types of layers: 1st convolution layers, 2nd pooling layers, and a 3rd classification layer. The 1st layers lie at the centre of the architecture. The

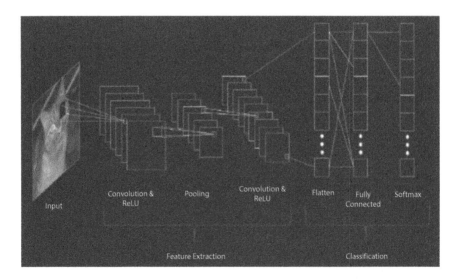

Figure 14.6 CNN.

loads characterize an intricacy functional to the 1ˢᵗ input, known to be the receptive field. The filtered resultant then undergoes through nonlinearity known as a feature map shown in Figure 14.6.

The 2ⁿᵈ layer is used to act as a non-linear down sampling with the trick of smearing a proportionate capacity, for instance, a component map comprising of the most non-overlapping subsets. Except when we decrease the size of component maps, the memory required by the layer prominently lessens the no of boundaries, resulting in over fitting. In the middle of convolution layers, these layers are embedded occasionally and afterward are presented as and inputs to expected DNN.

CNNs uses standard procedures that lessen over fitting. When we construct a model using dropout at the time of every construction lifespan, a predetermined position of nodes and their imminent and live activities are terminated. Including dropout commonly increases the accuracy and conclusive ability of an architecture as it progresses the prospect node [15].

14.2.1.6   Generative Adversarial Networks

Generative adversarial networks (GANs) are a type of architecture used in unassisted ML. These two neural networks are connected loosely to one another in such a way that each one wants to outmanoeuvre each other. Goodfellow *et al.* discovered a technique where one network plays the role of a generating block and another as a discriminative block. The generative block accepts input information and produces outputs with indistinguishable qualities as an actual input. On the other hand, a discriminative block accepts actual data and the data from the generative block and tries to identify if the provided data is genuine or counterfeit. When the drill

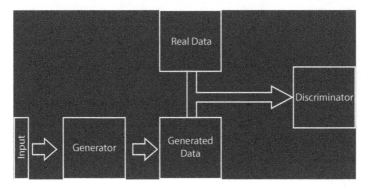

Figure 14.7 Generative adversarial networks.

is completed, the generative block is fully prepared for producing recent information that is not discernible from actual data shown in Figure 14.7.

As GANs demonstrated, an extensive range of applications are applicable through it, especially to pictures. Models incorporate picture upgrade subtitle age and visual stream approximation. In fact, Facebook has a pre-qualified, open foundational GAN for picture aging known as a deep convolutional generate ill-disposed system (DCGAN).

14.2.1.7 Recursive Neural Networks

A system that inputs a large of traffic load with a recursion to develop the data provided is known as a Recursive Neural System. The output in these systems works as a contributory input for the next step. At first, the initial two data sources are set as an input in the model. Then, the output obtained from that model is utilized as an input alongside the subsequent

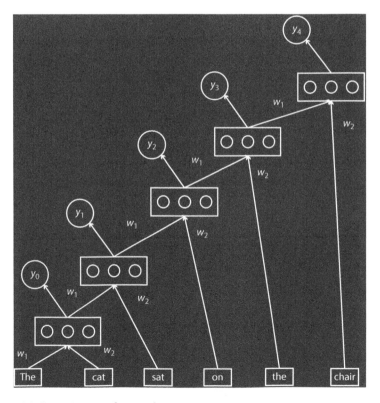

Figure 14.8 Recursive neural networks.

stage. This kind of model has been utilized for different common language handling undertakings and picture division shown in Figure 14.8 [16].

14.2.2   Applications of Deep Learning in Cyber Security

Since we secured probably the most extensively recognized dangers and digital assaults cyber security groups face, it is an ideal opportunity to clarify how Deep Learning applications can prove to be helpful.

14.2.2.1   Intrusion Detection and Prevention Systems (IDS/IPS)

IDS/IPS frameworks are used to distinguish within network protocols. They also keep interlopers away from getting the setup and restraining the client. To survive threats like data breaches, IDS/IPS use signature criteria and specific types of attacks they have excelled in.

Generally, IDS/IPS work under the strict roofs of Deep Learning algorithms. Be that as it may, these calculations made the framework produce some incorrect +ve, making security groups work generously and cause superfluous exhaustion.

DL, CNN, and RNN can be used to make more intelligent Intrusion Detection/Prevention Systems frameworks by probing the traffic with the most frequent accuracy, lessening the quantity of incorrect cautions, and serving refuge groups separate terrible and prodigious network happenings [17].

14.2.2.2   Dealing with Malware

Standard firewalls identify malware by employing a signature recognition system for conventional malware arrangements. For an organisation to develop a database for malware detection, it must refresh its database regularly to fuse new threats that were introduced recently. While these strategies are productive against recognised dangers, it combats to manage threats with new approaches or hacks.

Deep Learning does not depend on recollecting known signatures and common assault patterns; its calculations are equipped for recognizing further developed threats that are not dependent on it. Rather, they can build up their skills with framework and perceive dubious exercises that may show the immediacy of malware.

14.2.2.3   Spam and Social Engineering Detection

A Deep Learning method that works to distinguish and manage spam and different types of equipment used to build up social skills is known as Natural Language Processing (NLP). Natural Language Processing observes correspondence and linguistic patterns and uses different fact-based models to distinguish and restrict spam.

Two scientists, namely Tzotzil and Likes, conducted an experiment utilizing Deep Learning for organising junk messages, emails, etc. The objective of the experiment was to extricate topographies that rely on the body of an email and observe how many recurring words it might contain. To provide validation to the experiment, a DBN, with 3 buried layers from the unit lying in between an RBM, was used. The result of the experiment was that a DBN had a higher accuracy rate than SVM, varying with the following measurements: (DBN correctness) 99.45%, 97.5%, and 97.43% to (SVM exactness) 99.24%, 97.32%, and 96.92%.

14.2.2.4   Network Traffic Analysis

To dissociate HTTPS network traffic to obtain resultant search results for spiteful exercises, Deep Learning ANNs reflect reliable outcomes. It has an extreme value to manage numerous digital dangers, for example, SQL infusions and DOS attacks [18].

Wang conducted an experiment to use Deep Learning to identify the traffic type distinguishing proof. He used a stack autoencoders bridge with a sigmoid layer to receive classification results. Wang used TCP stream data as a dataset from an interior system and the payload bytes of every assembly. This dataset contains 58 various convention types, from which HTTP has been avoided on that level, that are anything but difficult to distinguish and present a vast dominant part of data [19]. After this process, a three-layer stacked autoencoder executes feature extraction on data and then, the output features are displayed to the sigmoid layer through which the process of classification takes place.

This system of characterisation displays accuracy ranging from 91.74%-100% and review ranging from 90.9%-100% depending on convention type.

14.2.2.5   User Behaviour Analytics

Any association that regularly deals with hundreds of users on a daily basis desperately needs to track and analyse client exercises as a significant safekeeping practice. User behaviour generally relies more on assumption testing than remarking conventional pernicious exercises against the systems as it sidesteps safety efforts and does not raise any alerts.

For example, when an insider threat occurs, admins use their authentic access for vindictive purposes. In such a case, invasion of the system is not occurring from outside, which reduces numerous digital barrier devices pointless in contradiction to such assaults.

User and Entity Behaviour Analytics (UEBA) are used as incredible equipment against such malicious acts. It can get ordinary worker standards of conduct and perceive dubious exercises, after an observational period, like getting access to the system in non-working hours, demonstrate a best example of an insider threat, and must immediately raise alarms and take other appropriate actions.

14.2.2.6   Insider Threat Detection

Theft to personal info, sabotage to system, etc. are considered to be major cyber security challenges and as we go deeper into their advancement, we can consider that the most dangerous threat in today's world is an insider threat. The incentives and patterns of insider threats are non-predictable and may vary widely from one another, even though the impairments caused are quite predictable and consistent. To provide a validated point of view to the theory, et al. used an unsubstantiated Deep Learning network on the framework for receiving filtered data through ordinary info. He tried two sorts of approaches: a DNN and an RNN. A component trajectory involving a synopsis of the framework datasets for every client was created for each day and offered as an input to a DNN/LSTM to obtain an output, respectively, for each client which will be considered as a next day's element vector for both DNN and RNN. At this point, when an output varies significantly from today's info, an inconsistency of data happens. Using a distinct setup for each client suggests the setup is not required to display the extensively profitable behaviour of the whole thing being similar. With DNN outdoing the LSTM, the models usually seem to perform well with the head segment examination and disengagement timberlands, an abnormality identification calculation comparative in development to irregular woods and setting the danger occasions over the 95th percentile of oddity scores [20].

14.2.2.7   Border Gateway Protocol Anomaly Detection

For automated frameworks, the Border Gateway Protocol (BGP) is a web convention that accounts for trading of guidance and reaches the time and

efficiency of data. The ability of BGP is very important for proper working of the web. Resolving the defects of BGP results into DDoS assaults,
snuffling, redirecting of paths, burglary of network topology information,
and so on. In this way, it is fundamental to distinguish irregular BGP occasions progressively to alleviate any possible harm. Cheng *et al.* extricated
thirty-three highlights from the information and utilized an LSTM, joined
with calculated relapse to recognize strange BGP traffic with 99.5% accuracy, a demonstrative enhancement over a Non-Deep Learning approach.

14.2.2.8   *Verification if Keystrokes were Typed by a Human*

Keystroke dynamics are a technique that use a biometric setup to collect the
timings of each candidate's keystroke with all its info. Kobojek and Saeed
discovered an idea to authenticate this experiment. Individuals submitted
their keystroke data with 12-25 examples for every individual and, by accu-
mulating noise to the human-created data, negative data was induced. This
data was then presented as an input to the RNN unit and then the sigmoid
unit for characterising data. In this process, GRU and LSTM methods were
both utilised. There are various restrictions in this trial, in light of the circumstances that the dataset was small in size and there was no signified
negative data, just some manipulated examples. Yet the inventors had the
luck to accomplish 80% exactness when there were no fake positive results
in an architecture comprising of two LSTMs and one sigmoid grouping
layer, exhibiting the potential to explore more.

14.3   Framework to Improve Security in Cloud Computing

We all are aware of how much physical security banks comprise of. Most
banks physically present will incorporate security highlights like surveillance cameras and impenetrable glass. Safety officers and bank workers
additionally help stop likely hoodlums and money is put away in exceptionally secure safes.

In any case, envision if, rather than being kept in one spot, each bank
office's money was put away in various safes everywhere throughout the
nation that were worked by an organization having some expertise in safe
upkeep. How could the bank be certain that its cash was secure without
conveying extra security assets around its dissipated safes? This is the thing
that cloud firewalls do shown in Figure 14.9.

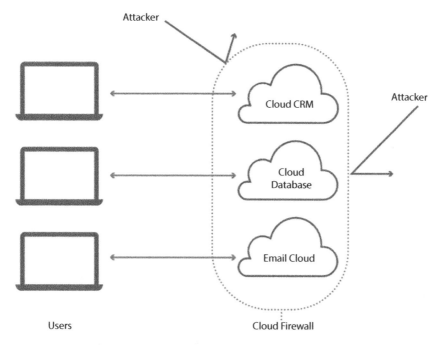

Figure 14.9 Security in cloud computing.

The cloud resembles a bank with dissipated assets, however, rather than cash, the cloud stores information and computational power. Approved clients can interact with the cloud from anyplace and on practically any system. Applications that run in the cloud can be running anyplace and that additionally applies to cloud platforms and frameworks.

Cloud firewalls block digital assaults coordinated at these cloud resources. As the name suggests, a cloud firewall is a firewall that is facilitated in the cloud. Cloud-based firewalls structure a virtual boundary around cloud stages, foundation, and applications, similar to a conventional firewall structure, as an obstruction around an association's inside system. Conveying a cloud firewall resembles supplanting a bank's local security cameras and a physical safety officer with a worldwide day in and day out security community that has brought together staff and surveillance cameras that take care of from all the spots where a bank's advantages are put away.

14.3.1   Introduction to Firewalls

A firewall is basically a safety framework that screens and regulates network traffic. Firewalls work in the middle of confided and malicious networks

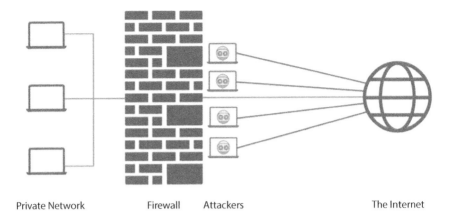

Private Network   Firewall   Attackers   The Internet

Figure 14.10 Firewall.

shown in Figure 14.10. For example, collage networks utilize firewalls as a shield to protect their system from working dangers.

Firewalls instruct whether it is really required to give permission for traffic to pass through it. They might be incorporated with equipment, programming source codes, or a mix. The term 'firewall' originated by observing the practice used to develop building walls in the middle or centre of any structures that contain fire [21].

14.3.2 Importance of Firewalls

The main motive to introduce firewalls to our life is a requirement of a secure environment. To ensure the safety of sensitive content, firewalls restrict and block the indicative traffic to approach the system.

Firewalls are used to filter content. For example, a college can recruit a firewall to keep students away from their system to keep them from getting to grown-up material. Additionally, in certain countries, the supervisors monitor a firewall that can forestall individuals inside a specific nation or state to get access to specific parts of the Internet.

Firewalls can be proven useful to accomplish many tasks. Some of them are mentioned below.

14.3.2.1 Prevents the Passage of Unwanted Content

- There is no restraint to dreadful and undesirable content over the web. Such undesirable content can get access to the framework unless a solid firewall is installed. A majority of the OS organizations of a firewall viably deal with undesired and harmful substances from the web.

- Whenever another framework is assigned any work, it must be checked by the client. If, by any chance, that firewall does not exist, at that point an outsider firewall can be introduced.

14.3.2.2   Prevents Unauthorized Remote Access

- Various abusive programmers exist today in the universe of computer networks and are hitting continuous attempts to gain access to susceptible systems. Some irresponsible clients are never cautious enough to realise who can get access to their system. A solid firewall forestalls such a chance of a forthcoming exploitative programmer getting portable admittance to a framework. Such distant access is unquestionably unapproved and might be suspected of damaging purposes as well.
- We require a strong firewall setup to ensure safety of information, transactions, etc. To undertake a spillage of private information and data implies enormous misfortune and disappointment. With reference to understanding the implication of firewalls to forestall unapproved isolated access, the examples of banking associations and national level security organizations are suited best.

14.3.2.3   Restrict Indecent Content

- The spider web of the Internet has exposed individuals, especially our youth, to malicious content. This web is spreading its claws very quickly. With drastic change in lifestyle patterns, this content has started playing with the innocent minds of individuals. Analysing these circumstances is very critical job with respect to caretakers to guarantee that such substances are forestalled in their PC frameworks.
- Introduction to such substances, including revolting ones, can be dangerous to youngsters' minds bringing about unusual practices and unethical leads. A strong firewall ensures PC frameworks by forestalling the passage of shameless and revolting substances and, along these lines, permits guardians to protect their children.

14.3.2.4   Guarantees Security Based on Protocol and IP Address

- Network Address Translation (NAT) is a type of firewall that successfully protects frameworks from harmful attackers that are not inside the range of network. Hence, the IP address of these systems can only be protected and visualised in their network.
- Equipment firewalls are valuable for analysing traffic exercises dependent on a specific convention. At whatever point a connection gets set up, since the earliest starting point is as far as possible, a pathway of training is reserved that contributes to keeping the system safe.

14.3.2.5   Protects Seamless Operations in Enterprises

- In today's world, many large-scale businesses, systems, and software enterprises have created dependencies for many associations. A regionalized DS (Distributed System) merges with the accessibility of information from any place over a whole geological nearness is empowering official investors to use and work on the information for fruitful corporate tasks.
- A client can log in to his system from any source using endorsements within the range of the network. It is quite elementary to have a strong firewall set up, holding such an enormous system framework and massive data. A firewall is the most significant part in conferring security to every one of these angles. It would be extremely hard for associations to have such consistent tasks and exercises would be severely hampered without using powerful firewalls.

14.3.2.6   Protects Conversations and Coordination Contents

- Third-party client interaction is the most essential part of the lifecycle of organizations that are in any way linked to service industries. They regularly share pertinent content with the client and internal teams as a basic part of various major and minor projects and not just data, but they also interact with companies as well as external investors through different types of virtual platforms.

- Almost all the content from these coordination activities is confidential and must be protected effectively and no organization can simply afford the cost of leakage of such important content. Firewalls monitor frameworks adequately and permit a secure and safe progression of data conferring a feeling of confidence to the stakeholders.

14.3.2.7   Restricts Online Videos and Games from Displaying Destructive Content

- In today's world, watching videos and playing online video-games has acquired a great space of our life. Many company websites and applications provide a virtual space for users to access, watch, and download such games and videos. Except some of the company websites, that does not ensure refuge of access and frequently, there is a massive difference in negative content as malware and viruses attempt to access the client's data. In such situations, there is a great requirement for a firewall in the framework as it keeps the client's system protected from all types of possible malware assaults through online games or videos.
- It is not easy to detect malware assaults, especially through websites that are providing access to online games and videos as clients are amped up for the movies and games that they supposed to examine over the web. Yet, it is better to get the system checked by a specialist to gain proper knowledge of whether the firewall is hardware or software, according to the needs of the system. Hence, if a client finds the urge to access an online video or game, they ought to adapt oneself with an indispensable firewall setup to make use of the firewall efficiently.

14.3.3   Types of Firewalls

14.3.3.1   Proxy-Based Firewalls

A proxy-based firewall is comparable to a watchman of a building or workspace. This watchman stops visitors before they enter the building to ensure they are not outsiders, outfitted, or in some other way a danger to the premise and people living there. The watchman also ensures that the known people have a safe method to return home and are not intending to drive under the influence.

The disadvantage of having a watchman at the building is that when many individuals are trying to enter or leave the building at the same time, there will be a long queue and some people will have delays. With reference to the above-mentioned situation, a significant disadvantage of a proxy-based firewall is it sources lifelessness, especially during times of heavy traffic.

A proxy is kind of a personal computer that works as a door between a big system and neighbourhood kind of arrangement, for example, the Internet. These are substitutions that works as a wall and lie in between the clients and servers. Clients connect to a computer that acts as an entrance between a local and larger network, such as the Internet. These are perfect examples of proxies that are situated in the middle of customers and workers. Customers interface with the firewall and the firewall assesses active bun Deep Learning, after which it will make an association with the proposed beneficiary (the web worker). Also, when the website worker sends a reaction to the customer, the firewall captures that request, evaluates the package, and later carries that reaction in a different association amid the customer and firewall. A proxy-based firewall adequately forestalls an immediate association between the customer and worker.

14.3.3.2 Stateful Firewalls

In SE, a stateful app standbys data from past occasions and associations in contradiction to it. A stateful firewall standbys data with respect to open associations and utilizes this data to break down incoming and outgoing traffic, as opposed to reviewing Deep Learning. These firewalls are quicker than intermediary based firewalls, as they do not examine each parcel. They require appropriate setting arrangements for making decisions. For example, if the firewall accounts for a departing packet requesting for an output, it might possibly permit approaching Deep Learning on that association in the event that they provide the requested kind of output it is looking for. These firewalls secure ports by keeping all their systems shut except in the case where they are approaching Deep Learning for packet demand access to a particular port. This process, known as port scanning, can easily moderate an attack. The only disadvantage of stateful firewalls is that anyone can take control of system by fooling a customer into mentioning a particular kind of data. When the customer demands that reaction, the aggressor would then be able to send vindictive parcels that coordinate with the standards through the firewall.

14.3.3.3   Next-Generation Firewalls (NGF)

NGF are the type of firewalls that comprise of the volume of conventional firewalls and additionally make use of a large group of added topographies to revert back to hazards in different levels of the OSI Model. Some NGFW unambiguous attractions are as follows:

- Deep Packet Inspection (DPI): Regardless of conventional firewalls, NGFs achieve meaningfully more inside and out of examination of parcels. This thoughtful valuation can visualise things like parcel cargos and which application is being grown to be Deep Learning. This licences the firewall to approve more coarse separating rules.
- App Cognizance: Permitting this element to work makes the firewall heedful of which apps are in working condition and which positions those apps are utilizing. This can protect against particular types of malware that propose to finish a seriatim process and subsequently accept a switch over its port.
- Distinctive Cognizance: This allows a firewall to implement rules that depend on character, for example, which client is signed in, which PC is being utilized, etc.
- Sandboxing: Firewalls can disconnect bits of code connected with imminent Deep Learning and implement them in a 'sandbox' sphere to assure they are not booming on malevolently. The repercussions of this sandbox test are able to be operated as trials when picking whether to let the parcels enter the system or not.

14.3.3.4   Web Application Firewalls (WAF)

While conservative firewalls help to safeguard private systems from harmful web apps, Web Application Firewalls provide assistance to protect web apps from spiteful people. A Web Application Firewall protects web apps by unravelling and detecting an HTTP stream of traffic between a web app and the Internet. It usually safeguards web apps from attacks like cross-site-scripting (XSS), SQL booster, file enclosure, and cross-site counterfeit shown in Figures 14.11, 14.12 and 14.13.

By transmission of a Web Application Firewall towards web application, a protective shield is placed between the web app and the Internet. While a deputation-based firewall guarantees a client machine's specifications by

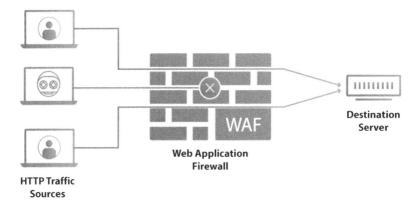

Figure 14.11 WAF.

operating a mid-Deep Learning person, a Web Application Firewall is a reverse-deputation protecting the worker from overview by making customers go through the Web Application Firewall before reaching the server.

A Web Application Firewall works under a lot of rules, habitually, called policies. These policies tend to form secure environments against possible threats in the application by separating out lethal traffic. The approximation of a Web Application Firewall comes to a level with the swiftness and ease with which strategy modification can be objectified, considering a faster response to updating all new kinds of attack vectors. At the time of a DDoS, cost restrictions can be executed instantly by regulating Web Application Firewall methods. Marketable Web Application Firewall items, like a cloud flare Web Application Firewall, protects a large amount of web apps from attacks every hour.

14.3.3.5   Working of WAF

A cloud flare WAF control panel permits users to construct influential rules through easy access and also delivers Terraform integration. Every entreaty to the WAF is reviewed against the risk intelligence and the rule locomotive created over twenty-five million websites. Mistrustful requests can be congested, tested, or recorded as per the needs of the user, while genuine requests reach directly to the end point. Analytics and cloud flare logs allow discernibility into unlawful metrics for the user.

14.3.3.6   How Web Application Firewalls (WAF) Work

WAF security shields web apps from unpleasant endpoints and basically works as contrary energy of delegation servers (for reverse proxies), which

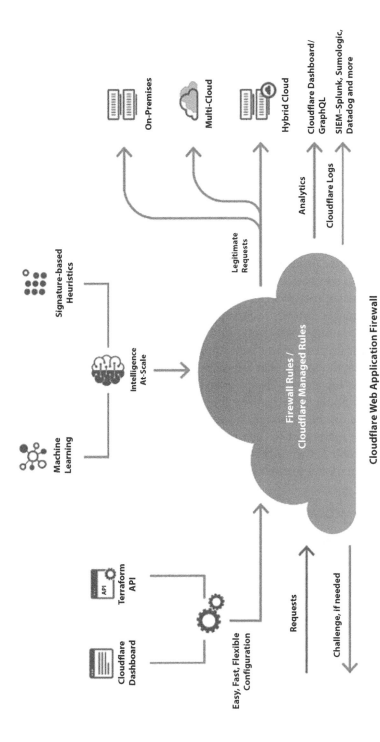

Figure 14.12 Working of WAF.

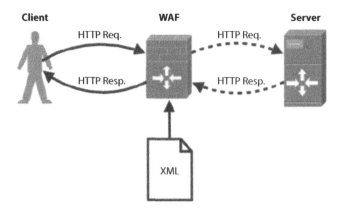

Figure 14.13 Cloud WAF.

defends gadgets from spiteful applications. Web Application Firewalls are developed as server-side software modules, hardware appliances, or FWaaS.

To assure security, WAFs restrict and examine all HTTP demands. Fake traffic is blocked or tries to be stopped with the help of CAPTCHA tests intended to stump unsafe bots and PC programs.

The delicate print of the Web Application Firewall association depends on safety methodology that are relies upon improved tactics, which must address the top most web app's security inadequacies chronicled by the Open Web Application Security Project (OWASP). Customarily, these terminologies can be thorough, necessitating particular heads to design the WAF as an agreement to the organization's security strategy. These heads are responsible for accurately placing, arranging, managing, and observing WAFs to guarantee greatest security.

14.3.3.7   Attacks that Web Application Firewalls Prevent

Web Application Firewall security can forestall numerous assaults, including:

- Cross-site Scripting: Attackers infuse client-side data into site blogs that can be easily accessed by different clients.
- SQL Infusion: Venomous code is infused into a web passage field that permits assaulters to make the app and hidden frameworks cost efficient.
- Cookie Theft: Tempering of a cookery to increase the level of unapproved data about the client to accomplish certain purposes, for example, identification theft.

- Invalidating Data: Attackers alter HTTP demand (counting the URL and structure fields) to sidestep the framework's security instruments.

14.3.3.8 Cloud WAF

A cloud WAF, also known as cloud based WAF, provides present day web apps security at a lower cost with much more than customary machine-based web application firewalls, while predefining some major points of interest. The customization and adaptability of this type of cloud Web Application Firewall administration spares executives from monotonous physical adjustment of security programming or equipment on their frameworks, takes into consideration proactive instead of responsive danger locations, empowers continuous application of security bits of knowledge and deceivability, and guarantees consistency, all while bringing together app security over multiple clouds, half on cloud, or on assumption based application situations.

Cloud based WAF management offer more flexible, adaptable, and responsive app security alternatives dependent on security approaches that have been predefined to scale and respond consistently to dangers per application or inhabitant.

14.4 WAF Deployment

1. Reverse Substitution
The Web Application Firewall is kind of a substitute to the app worker. In this way, gadget traffic goes straightforward from Deep Learning to the WAF.

2. Transparent Reverse Substitution
A Transparent Reverse Substitution is a converse substitution with a straightforward mode. Therefore, the Web Application Firewall independently sends parted traffic to web applications. Due to this, the IP envelope secures itself by concealing the location of the app worker. Execution is possible with the result being a drawback during interpretation if it is inactive.

3. Transparent Bridge
HTTP traffic shifts straightforward from Deep Learning to the web application.
This results in a WAF stuck between the gadget and the worker.

14.4.1   Web Application Firewall (WAF) Security Models

Web Application Firewalls consist of a positive security model and a negative security model or sometimes a mixture of both models. A positive security model WAF ("whitelist") rejects every operation which does not comprise of authorised permission, whereas a negative security model ("deny list") consists of a dropdown list of things that are restricted to be executed, providing permission to everything that is not part of that list.

Positive Web Application Firewalls as well as negative Web Application Firewalls security models both provide their inputs in many application security situations. For instance:

1. Positive Security Model
When we perform input validation, the positive Web Application Firewall model states that you restrict the filtering out of bad inputs in order to specify the allowed inputs. The importance of using a positive security model firewall is that new attacks, which are not anticipated by the developer, will be prevented.

When performing input approval, the positive model directs that you accept the permitted input instead of attempting to sift through bad input sources. The dominance of choosing a positive security model firewall is that new cyber traps, which are not foreseen by the engineer, will be pre-empted.

2. Negative Security Model
The negative Web Application Firewall model is simpler to execute, however, you can never be so certain that you have covered up all the aspects. You will certainly wind up with a considerable rundown of -ve marks to obstruct what must be perpetuated. The -ve security model methodology at first permits all traffic to come through despite the fact that many over the board restrictions have been implemented to improve security issues. Through this technique, offices that reliably make new system changes can get spare time so that the system does not get blocked.

14.4.2   Firewall-as-a-Service (FWaaS)

Firewall-as-a-Service, or FWaaS for short, is another term for cloud firewalls. Like other "as-a-Service" classes, for example, SaaS or PaaS, a FWaaS runs in the cloud and is spreading its roots all over the Internet and outside sellers offer them as a service that they update and maintain.

14.4.3   Basic Difference Between a Cloud Firewall and a Next-Generation Firewall (NGFW)

A cutting-edge firewall (NGFW) is a firewall that incorporates new advances that were not accessible in prior firewall items, for example:

Interruption Avoidance Framework (IPS): An interruption counteraction framework identifies and squares digital assaults.

Profound Bun Deep Learning Review (DPI): NGFWs investigate information parcel headers and payload, rather than simply the headers. This guides in recognizing malware and different sorts of malignant information.

Application Control: NGFWs can control what singular applications can access or square applications out and out.

NGFWs may have other propelled capacities too.

"Cutting edge firewall" is a comprehensively applied term, however, NGFWs do not really run in the cloud. A cloud-based firewall may have NGFW abilities, however, an on-premise firewall could likewise be an NGFW.

14.4.4   Introduction and Effects of Firewall Network Parameters on Cloud Computing

Secure Access Administration Edge, or SASE, is cloud-based systems administration engineering that consolidates organizing capacities, similar to programming characterized WANs, with a lot of security administrations, including Unlike conventional system administration models where the edge of on-premise server farms must be ensured with on-premise firewall, SASE has offered extensive security and access control at the system edge.

Inside the SASE organizing model, cloud-based firewalls work coupled with other security items to protect the system border from assaults, information breaks, and other digital dangers. Instead of utilizing numerous outsider sellers to convey and keep up each service, organizations can recruit a solitary merchant that packs FWaaS, cloud acquired security agents (CASB), secure web doors (SWG), and zero trust arrangements (ZTNA) with SD-WAN abilities.

14.5   Conclusion

Cyber Cloud Auscultation keeps on progressing at a cost, eclipsing digital safeguards capacity to compose and send new strategies to determine these upgraded assaults. This, joined with propulsion in machine learning calculation improvement, offers a great chance to work with neural network-based Deep Learning methods to deal with cloud digital indemnity applications to distinguish new variations of malware and zero-day mugging. In this chapter, we plotted the uses of Deep Learning strategies to a big miscellany of these cloud safety mugging types that focus on data, host systems, application software, and targeted networks. We likewise gave a thorough audit of the archived employments of Deep Learning strategies to recognize these digital assaults. We talked about the Deep Learning design and learned about well-known and developing techniques running from Recurrent Neural Networks to Generative Adversarial Networks. Current methodologies treat diverse assault types in disconnection. Future work ought to consider the falling association of malevolent exercises all through an attack lifespan (for example: penetration, abuse, order and control, data larceny, and so forth.). We likewise state certain measurements used to assess Deep Learning execution for digital security applications. In any case, the utilization of various datasets for preparing and testing did not take into account reasonable correlations over the entirety of the various methodologies. Accordingly, the requirement for point of reference datasets is basic for propelling Deep Learning in the digital security area. We recognized future observational openings identified with growing inaugurated datasets to galvanize work in growing new Deep Learning approaches for digital security and distinguishing the requirements for ways to deal with build-up that thinks about the foe, concerning how they can utilize Deep Learning as a device so that it can sabotage Deep Learning location mechanisms. This overview expects to give a helpful assortment of important stuff to propel analysts to propel the territory of Deep Learning for digital security setup.

References

1. Abhishek Kumar & Jyotir Moy Chatterjee & Pramod Singh Rathore, 2020. "Smartphone Confrontational Applications and Security Issues," International Journal of Risk and Contingency Management (IJRCM), IGI Global, vol. 9(2), pages 1-18, April.

2. Bhargava, N., Bhargava, R., Rathore, P. S., & Kumar, A. (2020). Texture Recognition Using Gabor Filter for Extracting Feature Vectors With the Regression Mining Algorithm. International Journal of Risk and Contingency Management (IJRCM), 9(3), 31-44. doi:10.4018/IJRCM.2020070103.

3. By Keith D. Foote :A Brief History of Deep Learning https://www.dataversity.net/brief-history-deep-learning/.

4. Cloud Flare https://www.cloudflare.com/learning/security/what-is-a-firewall/.

5. Daniel S. Berman, Anna L. Buczak *, Jeffrey S. Chavis and Cherita L. Corbett A Survey of Deep Learning Methods for Cyber Security.

6. David W Chadwick, Wenjun Fan, Gianpiero Costantino, Rogeriode Lemos, Francesco Di Cerbo, Ian Herwono, Mirko Manea, Paolo Mori, Ali Sajjad, Xiao-SiWang. A cloud-edge based data security architecture for sharing and analysing cyber threat information.

7. Dhirendra KR Shukla,Vijay KR Trivedi, Munesh C Trivedic Encryption algorithm in cloud computing.

8. Educba https://www.educba.com/firewall-uses/

9. Fatsuma Jauroac, Haruna Chiromab, Abdulsalam Y. Gitalc, Mubarak Almutairid, Shafi'i M. Abdul hamid, Jemal H. Abawajy Deep learning architectures in emerging cloud computing architectures: Recent development, challenges and next research trend.

10. John R. Vacca and Scott R. Ellis Firewalls: Jumpstart for Network and Systems Administrators.

11. Kumar Sahwal, ..., Kishore, ..., Singh Rathore, ..., & Moy Chatterjee, ... (2018). An Advance Approach of Looping Technique for Image Encryption Using in Commuted Concept of ECC. International Journal Of Recent Advances In Signal & Image Processing, 2(1)

12. Kumar, A., Chatterjee, J. M., & Díaz, V. G. (2020). A novel hybrid approach of svm combined with nlp and probabilistic neural network for email phishing. International Journal of Electrical and Computer Engineering, 10(1), 486.

13. Mohamed Amine Ferrag, Leandros Maglaras, Sotiris Moschoyiannis, HelgeJanicke Deep learning for cyber security intrusion detection: Approaches, datasets, and comparative study.

14. N. Bhargava, S. Dayma, A. Kumar and P. Singh, "An approach for classification using simple CART algorithm in WEKA," 2017 11th International Conference on Intelligent Systems and Control (ISCO), Coimbatore, 2017, pp. 212-216, doi: 10.1109/ISCO.2017.7855983.

15. NaLu, Ying Yang: Application of evolutionary algorithm in performance optimization of embedded network firewall.

16. Naveen Kumar, Prakarti Triwedi, Pramod Singh Rathore, "An Adaptive Approach for image adaptive watermarking using Elliptical curve cryptography (ECC)", First International Conference on Information Technology and Knowledge Management pp. 89–92, ISSN 2300-5963 ACSIS, Vol. 14 DOI: 10.15439/2018KM19

17. P. Ravi Kumar, P. Herbert Raj, P. Jelciana Exploring Data Security Issues and Solutions in Cloud Computing
18. Rathore, P.S., Chatterjee, J.M., Kumar, A. *et al.* Energy-efficient cluster head selection through relay approach for WSN. J Supercomputer (2021). https://doi.org/10.1007/s11227-020-03593-4
19. Saurabh Singh, Young-Sik Jeong, Jong Hyuk Park A survey on cloud computing security: Issues, threats, and solutions
20. Singh Rathore, P., Kumar, A., & Gracia-Diaz, V. (2020). A Holistic Methodology for Improved RFID Network Lifetime by Advanced Cluster Head Selection using Dragonfly Algorithm. International Journal Of Interactive Multimedia And Artificial Intelligence, 6 (Regular Issue), 8. http://doi.org/10.9781/ijimai.2020.05.003
21. Wikipedia https://en.wikipedia.org/wiki/Deep_learning Cloud WAF: Overview and Benefits https://www.globaldots.com/blog/cloud-waf-overview-benefits

About the Editors

Pramod Singh Rathore
Assistant Professor
Department of Computer Science & Engineering
Aryabhatta College of Engineering and Research Center,
Ajmer
Visiting Faculty, MDS University Ajmer, Rajasthan, India
pramodrathore88@gmail.com

Pramod Singh Rathore is pursuing his Doctorate in computer science from University of Engineering and Management (UEM) and done M. Tech in Computer Sci. & Engineering from Government engineering college Ajmer, Rajasthan Technical University, Kota India. He has been working as an Assistant professor of Computer Science & Engineering Department at Aryabhatta Engineering College and Research centre, Ajmer, Rajasthan and also Visiting faculty in Government MDS University Ajmer. He has total Academic teaching experience of more than 8 years with more than 50 publications in reputed, peer reviewed National and International Journals, books & Conferences like Wiley, IGI GLOBAL, Taylor & Francis Springer, Elsevier Science Direct, Annals of Computer Science, Poland, and IEEE. He has co-authored & edited many books with many reputed publisher like Wiley, CRC Press, USA. His research area includes NS2, Computer Network, Mining, and DBMS.

Dr. Vishal Dutt
Assistant Professor
Department of Computer Science
Aryabhatta College Ajmer, Rajasthan, India
Visiting Faculty, MDS University Ajmer, Rajasthan, India
vishaldutt53@gmail.com

Dr. Vishal Dutt is Doctorate in computer science from University of Madras, Chennai and has done MCA (Gold Medalist) from MDS University, Ajmer, Rajasthan, India. He has been working as the Assistant Professor of Computer Science at Aryabhatta College, Ajmer and also visiting faculty in Maharshi Dayanand Saraswati University (State Govt. University) Ajmer. He has total Academic teaching experience of more than 4 years. He has more than 35 publications in reputed, peer reviewed National and International, Scopus Journals & Conferences and Book Chapters. He has edited 2 books with Eureka publications. He has been keynote Speaker and resource person of many workshops and webinars in India. He has been the reviewer for Elsevier, Springer, and IEEE Access. He has been Program Committee Member and Reviewer in the International Conference on Computational Intelligence and Emerging Power System ICCIPS 2021. He has recently presented 2 articles in Sixth International Conference on Advances in Computing & Communication Engineering Las Vegas USA ICACCE 2020 (22-24 June) IEEE Explore Digital Library [SCOPUS] and 2 articles in the Third International Conference on Intelligent Communication Technologies and Virtual Mobile Networks (ICICV), 2021 (4-6 Feb. 2021) IEEE Explore Digital Library [SCOPUS]. His research area includes - Data Science, Data Mining, Machine Learning and Deep Learning. He also has Data Analytics Experience in Rapid Miner, Tableau, and WEKA. He has been working for more than 4 years in the field of data analytics, Java & Assembly Programming, Desktop Designing and Android Development.

Dr. Rashmi Agrawal
Professor
Manavrachna International Institute of Research and Studies, Faridabad, India
drrashmiagrawal78@gmail.com

Dr. Rashmi Agrawal is PhD and UGC-NET qualified with 18+ years of experience in teaching and research, working as Professor in Department of Computer Applications, Manav Rachna International Institute of Research and Studies, Faridabad, India. She is associated with various professional bodies in different capacity, life member of Computer Society of India, Senior member IEEE, ACM CSTA and senior member of Science and Engineering Institute (SCIEI). She is book series editor of Innovations in Big Data and Machine Learning, CRC Press, Taylor and Francis group, USA. She is the associate editor of Global Journal on Application of Data Science and Internet of Things. She has authored/co-authored many research papers in peer reviewed national/international journals and conferences. She has also edited/authored books with national/international publishers (IGI Global, Apple Academic Press, and CRC Press) and contributed chapters in books edited by Springer, IGI global, Elsevier and CRC Press. Currently she is guiding PhD scholars in Sentiment Analysis, Educational Data Mining, Internet of Things, Brain Computer Interface, Web Service Architecture and Natural language Processing. She is also an active reviewer and editorial board member in various journals.

Satya Murthy Sasubilli
Solution Architect
Huntington National Bank
satya.Murthy@huntington.com

Satya Murthy Sasubilli received his Post Graduation on Master of Computer Applications (M.C.A) from University of Madras, India in the year 2002. And he received his B.Sc. degree in Bachelor of Computer Science from Andhra University in the year 1999. He has more than 15 Years of experience in Cloud based Technologies like Big Data Solutions, Cloud Infrastructure, Digital Analytics delivery, Data Warehousing, Workday, In-Memory Technology, Cross-Channel Performance Evolution, Environmental trend analysis and predictive modeling for budget allocation and response forecasts.

Satya Murthy Sasubilli is currently working as a Solution Architect in Huntington National Bank, USA. He has been working with many Fortune 500 Organizations which includes Infosys, Capgemini, Mphasis an HP Company, Hexaware. As a solution Architect at Huntington National Bank he has helped multiple projects in successfully implementing Workday (Cloud). Satya is a regular contributor to National and International research platforms such as IEEE, SCRPOUS and SCIRP. He has been an active reviewer for Springer and Scientific Research (SCIRP) and EasyChair journals. Satya was a National Merit Scholarship winner in his School days and always excelled in STEM Courses. And he is an all-rounder in Cricket and did captaincy for his team and won many Domestic and District level matches.

Srinivasa Rao Swarna
Program Manager/Sr. Data Architect
Tata Consultancy Services
swarna.prince@gmail.com

Srinivasa Rao Swarna is currently working as a Program Manager/Sr. Data Architect in Tata Consultancy Services, USA. He has received his B.Tech degree in Chemical Engineering from Jawaharlal Nehru Technological University, Hyderabad, India and completed his internship at VOLKSWAGEN AG, Wolfsburg, Germany in 2004 in Simulation of used Motor Oils with Neural Networks by using MATLAB software. Since then, he has been working for many Fortune 500 Organizations, advising them in Data Privatization, Service Virtualization, Artificial Intelligence, Machine Learning, Big Data and Data Analytics across United States, Norway, Germany, and India for 16+ years. Srini is a regular contributor to National & International research platforms such as IEEE, SCRPOUS and SCRIP. He has been an active reviewer for Springer, Scientific Research (SCRIP) and Easy Chair Journals. Srini was a National Merit Scholarship winner in his school days and always excelled in STEM courses. He is an avid sports fan and follows Pittsburgh Steelers in NFL, Golf and Indian Cricket Team regularly.

Index

Also of Interest

Check out these other related titles from Scrivener Publishing

Also in the series, "Advances in Cyber Security"

CYBER SECURITY AND DIGITAL FORENSICS: Challenges and Future Trends, Edited by Mangesh M. Ghonge, Sabyasachi Pramanik, Ramchandra Mangrulkar, and Dac-Nhuong Le, ISBN: 9781119795636. Written and edited by a team of world renowned experts in the field, this groundbreaking new volume covers key technical topics and gives readers a comprehensive understanding of the latest research findings in cyber security and digital forensics. *EXPECTED IN EARLY 2022.*

Other related titles

SECURITY ISSUES AND PRIVACY CONCERNS IN INDUSTRY 4.0 APPLICATIONS, Edited by Shibin David, R. S. Anand, V. Jeyakrishnan, and M. Niranjanamurthy, ISBN: 9781119775621. Written and edited by a team of international experts, this is the most comprehensive and up-to-date coverage of the security and privacy issues surrounding Industry 4.0 applications, a must-have for any library. *NOW AVAILABLE!*

MACHINE LEARNING TECHNIQUES AND ANALYTICS FOR CLOUD SECURITY, Edited by Rajdeep Chakraborty, Anupam Ghosh and Jyotsna Kumar Mandal, ISBN: 9781119762256. This book covers new methods, surveys, case studies, and policy with almost all machine learning techniques and analytics for cloud security solutions. *NOW AVAILABLE!*

ARTIFICIAL INTELLIGENCE AND DATA MINING IN SECURITY FRAMEWORKS, Edited by Neeraj Bhargava, Ritu Bhargava, Pramod Singh Rathore, and Rashmi Agrawal, ISBN 9781119760405. Written and edited by a team of experts in the field, this outstanding new volume offers solutions to the problems of security, outlining the concepts behind allowing computers to learn from experience and understand the world in terms of a hierarchy of concepts. *NOW AVAILABLE!*

SECURITY DESIGNS FOR THE CLOUD, IOT AND SOCIAL NETWORKING, Edited by Dac-Nhuong Le, Chintin Bhatt and Mani Madhukar, ISBN: 9781119592266. The book provides cutting-edge research that delivers insights into the tools, opportunities, novel strategies, techniques, and challenges for handling security issues in cloud computing, Internet of Things and social networking. *NOW AVAILABLE!*

DESIGN AND ANALYSIS OF SECURITY PROTOCOLS FOR COMMUNICATION, Edited by Dinesh Goyal, S. Balamurugan, Sheng-Lung Peng and O.P. Verma, ISBN: 9781119555643. The book combines analysis and comparison of various security protocols such as HTTP, SMTP, RTP, RTCP, FTP, UDP for mobile or multimedia streaming security protocol. *NOW AVAILABLE!*